PRAISE FOR *SUSTAINABILITY IN TOURISM, HOSPITALITY AND EVENTS*

"This book is a golden resource for young THE professionals as they embark on their leadership journeys. The knowledge shared within these pages is the most valuable tool for the early stages of their careers. It will undoubtedly give them the preparedness for operating in this era of climate change. It is so comprehensive in covering the subject and provides multiple quality frame works to follow. Lera and Mason have created such a valuable tool."
Sue Williams, Positive Hospitality Ltd, UK

"Lera and Mason address a complex problem with clarity and candour. It is no mean feat to gather all of the factors that challenge us all in hospitality and put them into a concise and actionable fashion."
Sally Beck, General Manager, Royal Lancaster London, UK

"This is a very timely book, written in an accessible style, enriched with real-world examples, reflective questions and engaging activities. It is a valuable resource and introduction to sustainability in tourism, hospitality and events."
Alexander Trupp, Guest Professor, University of Innsbruck, Austria

"For anyone committed to the future of travel and hospitality, *Sustainability in Tourism, Hospitality and Events* by Lera and Mason is an essential resource. This guide skilfully combines useful theory with practical strategies, making the important idea of sustainability in tourism both clear and urgent. It goes beyond lofty ideals to offer a straightforward plan for adopting ethical and profitable practices in tourism. This book is not only important to read; it serves as an inspiring call to action that encourages readers to lead the way in creating a more responsible and resilient tourism industry for future generations."
Professor Walter Leal, Chair of Climate Change Management, Hamburg University of Applied Sciences, Germany

"A book that combines international real-world examples with the latest concepts and theories. It offers valuable insights not only for students but also for industry professionals seeking to update their knowledge and gain fresh perspectives on sustainability in tourism, hospitality and events."
Sandro Carnicelli, Professor of Tourism and Leisure Studies, University of the West of Scotland, UK

Sustainability in Tourism, Hospitality and Events

Fundamentals and practical approaches

Dimitri Lera and Michel Mason

> **Publisher's note**
> Every possible effort has been made to ensure that the information contained in this book is accurate at the time of going to press, and the publishers and authors cannot accept responsibility for any errors or omissions, however caused. No responsibility for loss or damage occasioned to any person acting, or refraining from action, as a result of the material in this publication can be accepted by the editor, the publisher or the author.

First published in Great Britain and the United States in 2026

All rights reserved. No part of this publication may be reproduced, stored in a retrieval system or transmitted in any form or by any means – including electronic, mechanical, photocopying, recording or by any artificial intelligence (AI) or machine learning system – without the prior written permission of the publisher. Unauthorized use, including the use of text or images to train AI models, is strictly prohibited and may result in legal action.

Kogan Page
Kogan Page Ltd, 2nd Floor, 45 Gee Street, London EC1V 3RS, United Kingdom
Kogan Page Inc, 8 W 38th Street, Suite 902, New York, NY 10018, USA
www.koganpage.com

EU Representative (GPSR)
Authorised Rep Compliance Ltd, Ground Floor, 71 Baggot Street Lower, Dublin D02 P593, Ireland
www.arccompliance.com

Kogan Page books are printed on paper from sustainable forests.

© Kogan Page 2026

The moral rights of the authors have been asserted in accordance with the Copyright, Designs and Patents Act 1988.

The content of this publication has not been approved by the United Nations and does not reflect the views of the United Nations or its officials or Member States.

www.un.org/sustainabledevelopment

ISBNs
Hardback	978 1 3986 2018 6
Paperback	978 1 3986 2017 9
Ebook	978 1 3986 2020 9

British Library Cataloguing-in-Publication Data
A CIP record for this book is available from the British Library.

Library of Congress Cataloging in Publication Data
A CIP record for this book is available from the Library of Congress.

Typeset by Integra Software Services, Pondicherry
Print production managed by Jellyfish
Printed and bound by CPI Group (UK) Ltd, Croydon CR0 4YY

*I would like to dedicate my first book to my parents Lilo and Lucio.
I hope I make you proud.
For Paulette, in loving memory.
I would like to thank my brother Marcantonio,
for his love and encouragement.*

*To the cherished memory of my dad Arthur,
and Paul, the smile in my heart.
To my mum Valerie and sister Nicol, thank
you for your patience and understanding.*

CONTENTS

List of figures and tables xi
About the authors xiii
Foreword xiv
Acknowledgements xvi
Walkthrough of textbook features and online resources xviii

1 **Introduction to sustainability in the tourism, hospitality and events industry** 1
   Tourism, hospitality and events sustainability in context 1
   About this book 4
   Part One: Framing sustainability for a changing industry 5
   Part Two: Applying sustainability in professional practice 7
   Key features of this book 10
   Student voice 13
   References 15

PART ONE Framing sustainability for a changing industry

2 **Sustainability in the global context** 19
   Introduction 21
   1972: Stockholm Conference and United Nations Environment Programme (UNEP) 22
   1975: The UN World Tourism Organization (UNWTO) 22
   1987: The Brundtland Report: *Our Common Future* 23
   1989: The Hague Declaration on Tourism 23
   1992: Earth Summit and Agenda 21 24
   1994: Elkington's Triple Bottom Line (TBL) 24
   2001–05: Cultural diversity and the role of Tourism, Hospitality and Events (THE) in sustainable development 26
   2010: The pillars of sustainability 28
   2015: The 2030 Agenda for Sustainable Development and the Sustainable Development Goals (SDGs) 34
   2017: Key competencies for sustainability 42
   2018–26: Next Tourism Generation Alliance and Pact for Next Tourism Generation Skills (PANTOUR) 43

2021–25: EU Transition Pathway for Tourism and the Sustainable EU Tourism project 44
2022: GreenComp, The European Sustainability Competence Framework 45
2024: The Statistical Framework for Measuring the Sustainability of Tourism (SF-MST) 49
Conclusion 49
Further resources 51
References 52

3 Professional development for sustainability 55

Introduction 57
Developing a sustainability mindset 58
SDG 4: Quality Education, Lifelong Learning and Continuous Professional Development (CPD) 59
Grounding sustainability in professional development 61
Key competencies for sustainability 62
Extending competency through the OECD Learning Compass 66
Designing effective professional development 70
Anderson's five Ps of professional development 71
Training needs analysis for sustainability in THE 78
Conclusion 84
References 86

4 Responsible practice for cultural sustainability 88

Introduction 90
Understanding cultural heritage 91
International frameworks for cultural sustainability 96
Challenges to cultural sustainability 99
Overtourism and its impact on cultural heritage 103
Corporate social responsibility and stakeholder theory 112
Designing ethical cultural experiences: Aligning practice with principle 117
Aligning cultural sustainability with the Sustainable Development Goals (SDGs) 119
Implications for the future of cultural tourism 123
Conclusion 124
Further resources 126
References 126

PART TWO Applying sustainability in professional practice

5 Measuring impact and reporting 131
Introduction 132
Ethical business practices 136
Corporate social responsibility (CSR) 137
Introduction to Environment, Social and Governance (ESG) 140
Carbon footprint and Scope 1, 2 and 3 emissions 144
Introduction to sustainability audits 147
Industry tips for reducing carbon footprint 151
International certifications: Validating sustainability efforts 156
Conclusion 164
Further resources 165
References 166

6 The circular economy 169
Introduction 170
Greenhouse gases (GHGs) and net zero 171
Integrating circular economy principles in THE 172
Sustainability in the circular economy 181
Cross-industry collaboration and circularity 188
Conclusion 195
References 196

7 Food waste 199
Introduction 200
Definition of food waste 201
Food waste: The scale of a global issue 201
Global impact of food waste on natural resources 205
The food waste situation in the UK 208
THE food waste 212
Food waste reduction interventions 217
The human factor in food waste reduction intervention 222
Conclusion 224
Further resources 227
References 227

8 Basics of sustainable menus: Principles and practices 230
Introduction 231
The role of the menu 232

Considerations for a sustainable menu 234
Menu layout 240
Food miles (or kilometres) 241
The Planetary Health Diet 245
The three-step sustainable menu: Planning, Designing, Presenting (PDP) 246
Bringing the sustainable menu to life: Operations 248
Networks 251
Conclusion 253
Further resources 256
References 256

9 Greenwashing 260
Introduction 261
The risks of greenwashing 262
Regulatory bodies 263
The seven sins of greenwashing 268
From compliance to accountability: Regulation, litigation and the future of credible sustainability 269
Greenwashing in THE operations 275
Avoiding instances of greenwashing 279
The danger of greenhushing 281
Conclusion 282
Further resources 284
References 285

10 Conclusion 287
Sustainability as practice in THE 287
Framing the professional landscape 287
Interpreting practice in context 292
Moving forward 295
An invitation to ongoing practice 297

11 Extension material: Real-world examples and professional interviews 299
Introduction: Purpose and use of extension material 299
Extended real-world examples 300
Professional interviews 306
Concluding reflection 313
Further resources 314
References 314

Index 316

LIST OF FIGURES AND TABLES

Figures

Figure 2.1	The four pillars of sustainability 29
Figure 2.2	The Sustainable Development Goals with the UN emblem 37
Figure 2.3	SDG 4 Quality Education 39
Figure 3.1	Cross-sector relevance of SDG 4.7.1 60
Figure 3.2	Anderson's five Ps framework for effective professional development 71
Figure 3.3	Applying Anderson's concept of Purpose in the five Ps to professional development in a THE context 73
Figure 3.4	Extract of the application of Purpose to professional development training when reading down a column 74
Figure 3.5	Extract of the application of Purpose to professional development training when reading down across two columns 74
Figure 3.6	Alzahmi and Alshamsi's three-tiered approach to training needs analysis 79
Figure 3.7	Top-down approach for THE to shift to a sustainable tourism mindset 80
Figure 3.8	Bottom-up approach whereby the individual initiates PD for a sustainable tourism mindset 81
Figure 3.9	Green Pearls® 82
Figure 4.1	The adoption of the Mediterranean diet on the Representative List of the Intangible Cultural Heritage of Humanity 95
Figure 4.2	The dilemma of the business of tourism and cultural appropriation 100
Figure 4.3	International tourist arrivals by region (2010, 2020, 2030) 104
Figure 4.4	The IceHotel, Jukkasjärvi, Sweden 116
Figure 4.5	SDG 4 119
Figure 4.6	SDG 8 120
Figure 4.7	SDG 10 120
Figure 4.8	SDG 11 121
Figure 4.9	SDG 12 121
Figure 4.10	SDG 16 122
Figure 4.11	SDG 17 122
Figure 5.1	CSR vs ESG differences 133
Figure 5.2	The Statistical Framework for Measuring the Sustainability of Tourism (SF-MST) 135

Figure 5.3	Scope 1, 2 and 3 emissions 144
Figure 5.4	A five-step approach to sustainability auditing 150
Figure 5.5	The process of obtaining international certifications 159
Figure 6.1	The carbon footprint of global tourism 172
Figure 6.2	The linear economy cycle 173
Figure 6.3	Linear economy cycle assumption 173
Figure 6.4	The 8Rs framework 175
Figure 6.5	A waste hierarchy model 184
Figure 6.6	Sustainability initiatives undertaken by the organizers of Glastonbury Festival 190
Figure 7.1	Total carbon footprint 202
Figure 7.2	Food waste along the supply chain 208
Figure 7.3	THE in the food value chain 213
Figure 8.1	The Planetary Health Diet 245
Figure 9.1	List of UK regulators 266

Tables

Table 2.1	The alignment of GreenComp competences with UNESCO's key competencies for sustainability 46
Table 2.2	Timeline of key milestones in global sustainability and their relevance to THE 47
Table 7.1	Nine influences on menu planning 218
Table 7.2	12 stages of menu design 219
Table 8.1	The environmental benefits of sustainable menu items 236
Table 8.2	Calculating food miles for a sustainable menu 242
Table 8.3	Sustainable menu checklist 252
Table 9.1	Three examples of greenwashing pitfalls 268
Table 9.2	The seven sins of greenwashing 270
Table 9.3	Patchwork of greenwashing regulation 271

ABOUT THE AUTHORS

Dimitri Lera is a specialist in sustainable tourism, hospitality and events, bringing together over 30 years of international managerial experience with a strong academic foundation. He was a lecturer in Sustainable Tourism, Hospitality and Events at the Edge Hotel School, University of Essex, where he led curriculum innovation focused on embedding sustainability into vocational higher education. He is currently completing a PhD investigating practice-based pedagogy for sustainability education in applied hospitality contexts. His work was awarded with the CHME Annual Award for Excellence in Teaching and Learning, the Greengage Sustainability Award for his innovative module design, and the Excellence Educator Award at the University of Essex. He is a UK ambassador for the Guardians of Grub campaign, dedicated to reducing food waste in the hospitality sector, and a UK member of the Slow Food movement. He has presented at major conferences including the Council for Hospitality Management Education (CHME), the Association for Events Management Education (AEME), the European School of Sustainability Science and Research (ESSSR) and the Advance HE Teaching and Learning Conference (Advance HE). He was also a panellist at the Green Event Innovations Conference 16 (GEI16). His work bridges industry relevance and academic insight to help shape a more sustainable future for the visitor economy.

Michel Mason is a sustainability educator and academic collaborator who specializes in embedding sustainable practice across higher education, As former Sustainability Engagement Manager at the University of Essex, she led a wide range of initiatives to embed sustainability into teaching, research and staff development. Her work involved close collaboration with academic departments, student services and professional staff across the university. She worked extensively with the Essex Business School and Edge Hotel School to support staff and students in embedding sustainability into curriculum design, experiential real-world projects and professional practice.

Her work focuses on practical strategies for curriculum transformation, institutional engagement and educator development, contributing to measurable improvements in national sustainability rankings and long-term cultural change within higher education. She has presented at major conferences and contributed to national and international dialogues on education for sustainable development through publications, academic networks and outreach.

She is currently completing a PhD that explores how sustainability can be more effectively embedded into teacher education and wider higher education contexts. She brings an interdisciplinary approach to her work, shaped by broad experience across academic and professional environments.

FOREWORD

As sustainability continues to evolve as a critical concern across global industries, the tourism, hospitality and events (THE) sectors find themselves at the centre of important conversations about environmental responsibility, ethical practice and long-term resilience. While these sectors are deeply interconnected and often operate in overlapping ways, they also represent distinct fields with their own histories, professional standards and academic trajectories.

This book brings these sectors together under the shared acronym THE, not simply for convenience but as a deliberate and meaningful choice. The decision to use THE reflects our aim to bridge varied perspectives while acknowledging the unique identities of each sector. Many students, for example, are enrolled in dedicated programmes in hospitality or events management, where the sustainability issues explored in this book are directly relevant to their chosen disciplines and career paths. At the same time, we recognize that some readers may come from backgrounds where tourism is used as an umbrella term encompassing all three areas.

Sustainability challenges cut across traditional boundaries. By framing tourism, hospitality and events together, we aim to help readers understand the need for a shared approach to problem-solving and to highlight the value of collaboration in addressing complex, interdependent issues.

> **INDUSTRY VOICE: KATE NICHOLLS OBE**
>
> Kate Nicholls OBE, CEO of UKHospitality, says: 'Sustainability in hospitality isn't just about cutting carbon – it's about securing the future of the whole supply chain, from farm to fork. Hospitality leaders must continue to collaborate with producers and policymakers to shape effective land use, supply resilience, and investment in green innovation. This is not a fringe issue – it's central to the sustainability agenda of the sector.'

In adopting THE as a collective term, this book encourages readers to consider sustainability as a sector-wide responsibility, while remaining attentive to the distinctive roles and contributions of each profession. It supports an integrated, cross-sectoral mindset, one that is increasingly necessary for navigating the demands of sustainable development in an evolving global THE context.

This book is designed for a diverse learning community: students preparing for careers in THE, educators shaping sustainability-focused curricula and new professionals seeking to navigate the demands of working in THE sustainably. It offers a foundation of understanding and a set of applied resources that support critical inquiry, strategic thinking and reflective professional development.

While the book aims to provide a comprehensive foundation for understanding sustainability in the tourism, hospitality and events sectors, it does not attempt to address every issue in equal depth. Certain topics that are critically important – particularly those relating to social justice, human rights and equity in the workplace – are not explored in full here. This is not a reflection of their lack of relevance, but rather a consequence of the book's focus and structure. Our views on sustainability are more critical, but this book is designed as a practical tool for the readers, not a platform for debate. Our intent is to equip readers with core concepts and the analytical tools needed to support meaningful engagement with sustainability in practice.

We encourage further reading and critical exploration beyond this text, recognizing that sustainability intersects with a wide range of social, economic and ethical concerns.

The content is designed to build both insight and capability supporting readers to make sense of sustainability not as an abstract ideal but as a practical and evolving part of professional life in THE. Through critical discussion, real-world examples and reflective activities, the chapters encourage readers to engage with the complexities of sustainability at multiple levels: from international policy to local practice, from organizational culture to individual decision-making. The book aims to strengthen the skills, values and judgement needed to respond to sustainability challenges with confidence and creativity, wherever readers find themselves in the sector.

The challenges ahead are significant but so too is the potential for positive change. With curiosity, commitment and collaboration, those working in tourism, hospitality and events can be powerful agents of sustainable transformation.

> **INDUSTRY VOICE: WALTER LEAL FILHO**
>
> Walter Leal Filho, Professor, HAW Hamburg and Manchester Metropolitan University, says: 'The transformative nature of sustainability-oriented networks may be better understood, if it is considered that they not only provide information, but also act as platforms to share best practices, develop new approaches, and build capacity.'

ACKNOWLEDGEMENTS

We extend our heartfelt thanks to the diverse range of contributors and collaborators who have enriched this textbook with their insight, experience and passion. This work has been shaped by voices from across the tourism, hospitality, events and sustainability sectors, including leaders from industry, respected academics and the next generation of professionals, our students.

We are especially grateful to those who generously shared real-world examples, expert commentary and thought-provoking interviews. Their contributions have brought authenticity and depth to this work, ensuring it reflects the realities and future of sustainable practice in tourism, hospitality and events.

With appreciation, we acknowledge the following individuals, listed in alphabetical order, for their invaluable support and knowledge sharing:

Kaitlin Arens, resort experience ambassador, Edgewood Tahoe Resort

Sally Beck, general manager, Royal Lancaster London

Mara Biebow, communication manager, Green Pearls

David Bradbury, environmental sustainability

Nicola Cade, lecturer in tourism and events, Edge Hotel School, University of Essex

Molly Eccleston, Head Start apprentice, Hallgarten and Novum Wines

Adrian Ellis, director, Hospitality Connect

Kate Emmett, hospitality sustainability consultant and head of carbon reduction, Zero Carbon Services

Steven England, director and principal consultant facilitator, The Art of Sustainability

Susan Furber, Kogan Page editor, for her invaluable advice and patient support with the writing of our first book

Walter Leal Filho, Professor of Climate Change Management at HAW Hamburg, Professor of Environment and Technology at Manchester Metropolitan University

Rowen Halstead, food waste advocate and private chef consultant

Salvador Holstein, director, Hotel Casa Palmela

Banthita Hunt, lecturer, Edge Hotel School, University of Essex

Kate Nicholls OBE, CEO, UKHospitality

Kai Parfitt, director of ESG, risk and compliance, Maybourne

Lynsey Penny, sustainability sector specialist, Guardians of Grub

Miranda Simmons, production executive, sustainability trainer

Whitney Vernes, Lecturer, Edge Hotel School, University of Essex

Sue Williams, hotelier and independent sustainability advisor

Andrea Zick, PA and chair of the Harvey Nichols Sustainability Forum

WALKTHROUGH OF TEXTBOOK FEATURES AND ONLINE RESOURCES

Chapter aim

Highlights for you the main issues and topics that will be covered in each chapter.

> **CHAPTER AIM**
>
> The aim of this chapter is to examine the evolution of sustainability within the global tourism, hospitality and events (THE) sectors, as shaped by international policy and governance. It traces key historical and contemporary developments led by the United Nations (UN), the UN World Tourism Organization (now UN Tourism), and the European Commission which have positioned sustainability as a core principle in THE.

Learning outcomes

Summarize what you can expect to learn, to help you track your progress.

> **LEARNING OUTCOMES**
>
> Upon completion of this chapter, you will be able to:
>
> - explain the historical development of sustainability in the global context of tourism, hospitality and events
> - demonstrate critical thinking competency by identifying and evaluating the influence of international institutions such as the UN, UN Tourism and the European Commission on sustainability policies and practices in THE sector
> - apply the four pillars of sustainability (economic, environmental, social and cultural) to a real-world THE example using systems thinking and critical thinking competencies

Real-world examples

Examples to illustrate how key ideas and theories are operating in practice to help you place the concepts discussed in real-life contexts.

REAL-WORLD EXAMPLE Kaikōura, New Zealand

Background: Kaikōura, located on the east coast of New Zealand's South Island, is a small coastal town internationally renowned for marine-based tourism, particularly whale watching.

Reflective questions

Questions to challenge your understanding of key concepts and apply theory to practice.

REFLECTIVE QUESTIONS

1 What is a sustainable menu and why is it important in the food and beverage industry?
2 How can a menu influence consumer choice and promote sustainability?

Activities

Activities to encourage you to reflect on what you have learned and apply your knowledge and skills in practice.

ACTIVITY 2: APPLYING SDG 12 IN THE

Background

SDG 12 Responsible Consumption and Production aims to 'ensure sustainable consumption and production patterns' (UN, 2015).

Key words

Important terms and concepts are highlighted at the start or throughout the chapter. Understanding these key words helps you to build subject-specific vocabulary and improves your ability to engage with academic texts and professional discussions in your reading.

> **KEY WORDS**
>
> Sustainable THE, THE governance, international frameworks, policy development, cultural heritage, sustainable development goals, competence frameworks, next tourism generation.

Key takeaways

Summarize the key themes for you to synthesize your learning.

> **KEY TAKEAWAYS**
>
> - Understanding sustainable menus:
>   - A sustainable menu involves promoting environmentally friendly and sustainable food choices.
>   - It is crucial to recognize the importance of sustainable menu design as it not only supports environmental stewardship, but also aligns with economic viability and social responsibility.
>   - It is key to contributing to long-term profitability and customer retention.

Additional learning features

Each chapter also highlights industry voices, key competencies and the relevant Sustainable Development Goals (SDGs) to help you apply theory to real-world practice for your future career.

Further resources

Links to continue your research and understanding.

References

Detailed references to provide quick and easy access to the research and underpinning sources behind the chapter.

Online resources

This book includes online resources for you comprising:

- chapter slides
- real-world examples
- activity answer keys

These resources can be accessed through the Kogan Page website: www.koganpage.com/STHE

1 Introduction to sustainability in the tourism, hospitality and events industry

Tourism, hospitality and events sustainability in context

The tourism, hospitality and events (THE) sectors became increasingly significant from the middle of the 20th century onwards. In the context of post-war reconstruction and economic liberalization, national governments across a range of economies identified tourism as a strategic mechanism for development. Tourism offered access to foreign exchange earnings and was widely promoted as a stimulus for employment, infrastructure investment and regional regeneration (UN Tourism, .n.d.). As cross-border travel expanded, and leisure time became more widely available, new destinations were developed and promotional efforts scaled up to attract international visitors.

In this phase, tourism was closely tied to modernization narratives. Governments positioned tourism as a symbol of progress and economic dynamism, often supported by large-scale infrastructure programmes. International air routes were extended, road networks upgraded and accommodation sectors formalized. Events also became tools for place-branding while hospitality expanded to meet the needs of growing middle-class markets. Over time, tourism, hospitality and events became embedded in urban planning, trade negotiations and development policy.

Yet from the 1970s, questions began to emerge about the long-term consequences of this growth. While economic benefits were measurable, they were not evenly distributed. In many places, the resources required to support tourism including land, water and labour were drawn from communities whose access to decision-making was limited. Housing, transport and public services were often reoriented around the needs of visitors. In parallel, ecological degradation became increasingly visible. Fragile ecosystems were altered to accommodate resort development, while emissions from aviation and energy use began to attract critical attention. These concerns contributed to the establishment of multilateral mechanisms aimed at improving global environmental governance, including the founding of the World Tourism Organization in 1975, later rebranded as UN Tourism (UN Tourism, .n.d.).

The period from the 1980s to the early 2000s was marked by the formalization of sustainability in international policy. Following the publication of the Brundtland Report in 1987, sustainable development was defined as 'meeting the needs of the present without compromising the ability of future generations to meet their own needs' (World Commission on Environment and Development, 1987). This definition, while deliberately broad, introduced a framework in which environmental protection, economic viability and social equity were understood as interdependent. These ideas were reinforced by Agenda 21, adopted at the 1992 Earth Summit, which specifically highlighted THE as a sector requiring better regulation and more equitable forms of development (UN, 1992).

Within THE, these policy shifts prompted new efforts to incorporate environmental criteria into planning, certification and strategy. National governments began to publish tourism development plans that referenced environmental safeguards and stakeholder consultation. At the same time, intergovernmental bodies and non-governmental organizations developed voluntary guidelines intended to support sustainable practice. Sustainability was increasingly treated not as a constraint but as a set of performance standards against which destinations and operators might be assessed.

These developments reflected growing concern about the pressures placed on local environments and social systems. Resource use in tourism-related infrastructure frequently exceeded what ecosystems or communities could absorb. Emissions from long-distance travel continued to rise (UNEP, 2024). In areas where economic dependency on tourism increased, concerns grew about volatility, seasonality and the ability of host communities to cope with extrenal disruptions such as shifts in international travel demand or changes in foreign market conditions (UN, 2015). Sustainability in this context was framed not only as a matter of environmental protection, but also as a question of ethical responsibility, long-term viability and institutional reform.

As sustainability frameworks matured, they were incorporated into broader development agendas. The launch of the Sustainable Development Goals in 2015 established an explicit expectation that economic sectors, including THE contribute to objectives such as climate action, decent work and reduced inequalities (UN, 2015). The goals also encouraged a shift from mitigation to transformation. This meant going beyond technical improvements to interrogate how decisions were made, who benefitted from them, and what assumptions guided the allocation of resources and responsibilities.

In practice, responses to these challenges have been uneven. Some progress has been made in areas such as energy efficiency, responsible sourcing and environmental education. However, implementation has often been selective or limited to isolated initiatives. Many businesses continue to rely on resource-intensive practices, even as they adopt sustainability rhetoric. Emissions remain concentrated in high-consumption segments of the market. The structural features that produce inequality such as fragmented labour conditions, uneven access to land and limited stakeholder participation have proven difficult to address through voluntary codes or market incentives alone.

According to the World Travel & Tourism Council (WTTC, .n.d.), in 2024 THE sectors generated $10.9 trillion in global GDP, accounting for 10 per cent of global economic activity. These sectors supported 357 million jobs worldwide, equal to one in every ten workers. Visitor spending totalled $7.2 trillion, with international travel contributing $1.9 trillion and domestic travel $5.3 trillion. Projections indicate continued growth, with an expected contribution of $16.5 trillion by 2035 and more than 460 million jobs (WTTC, 2024).

This scale of activity has implications for global emissions. Oxford Economics (.n.d.) estimates that in 2022, tourism alone was responsible for 5.3 per cent of total greenhouse gas emissions. These emissions derive not only from aviation and road transport but also from accommodation, construction, supply chains and waste. Tourism-related food systems, including international food transport and high-volume production, contribute significantly to the environmental footprint of the sector (Yu et al, 2025). Without structural changes, these impacts are projected to intensify.

Labour issues further complicate the sustainability landscape. In many parts of the world, work in THE is precarious, low paid and lacking in progression opportunities. Skills development is inconsistent, and retention is hindered by short-term contracts and poor conditions. Industry research shows that a significant proportion of workers leave their positions within the first three months, citing lack of support and mismatched expectations (Gabbitas, 2023). These patterns persist despite growing acknowledgement that a stable, skilled and well-supported workforce is essential to delivering more sustainable outcomes (UN, 2015).

The challenges extend to the cultural dimensions of THE. While the sectors often promote their role in supporting cultural expression and heritage, the effects are mixed. In some contexts, commercialization has contributed to the simplification of cultural practices, raising concerns about representation, control and authenticity. The balance between creating economic opportunity and protecting the integrity of cultural forms remains difficult to navigate. Sustainability in this domain requires more than financial investment; it depends on careful negotiation of power, identity and meaning over time.

Recent years have seen growing interest in the circular economy as a model for reducing environmental pressure. This approach shifts focus from linear systems of production and disposal to practices of reuse, repair and regeneration (Ellen MacArthur Foundation, 2021). Within THE this has included experimentation with material recovery, energy recirculation and closed-loop sourcing. However, circular approaches remain the exception rather than the norm, and their scalability in large, dispersed service industries is still being evaluated.

The issues outlined here reflect persistent tensions in how sustainability is framed and pursued. These tensions are not easily resolved. They are embedded in infrastructure, finance, governance and the everyday practices of organizations and consumers. Addressing them requires more than individual action or technical innovation. It involves rethinking how decisions are made, how value is measured and how responsibility is distributed across institutions and geographies.

This book is situated within that context. It responds to the need for a more integrated understanding of the tourism, hospitality and events industries – one that recognizes both the scale of their contributions and the depth of their challenges. The sectors have evolved in tandem with global patterns of economic growth, mobility and consumption. Their future will depend on how well they are able to respond to changing expectations, ecological limits and the demands of more inclusive and resilient development. This book provides a foundation for that conversation.

About this book

This book has been developed in response to the growing need for a more coherent and critically informed approach to sustainability in the THE sectors. The sustainability challenges facing these industries are systemic in nature and cannot be adequately addressed through isolated interventions or surface-level change. Effective responses require an understanding of the historical dynamics that have shaped current practices, the policy frameworks that guide contemporary industry action, and the competencies needed to embed sustainability meaningfully within diverse organizational contexts.

Reflecting the multidisciplinary and professionally diverse nature of THE, this book is intended for:

- undergraduate and postgraduate students who require both theoretical and applied learning opportunities
- early career professionals entering or repositioning themselves within the sector and seeking to develop the knowledge, skills, attitudes and values necessary to respond to evolving sustainability expectations
- lecturers and educators who can use this book to structure curricula, facilitate critical engagement with key topics and support analysis of current challenges and future scenarios

Alongside technical skills and systems thinking, this book highlights the importance of self-awareness, reflective practice and ethical sensitivity as essential dimensions of sustainability.

Recognizing the complexity inherent in sustainability where business systems, human behaviour and socio-ecological dynamics intersect, this book does not claim to offer a singular solution. A one-size-fits-all approach is not only unrealistic but potentially counterproductive in such a varied and evolving field. This book does not aim to instruct, but to empower. It provides tools for questioning assumptions, co-creating knowledge and cultivating the critical judgment needed for sustainability to become a lived, not just learned, practice.

Rather than offering universal solutions, the book recognizes that sustainability is always situated. Readers are encouraged to explore how their personal, cultural and institutional contexts shape the challenges and opportunities they face. The book provides a framework for critical inquiry, alongside practical examples that support readers in their individual and collective journeys toward more sustainable practices, both professionally and personally.

To support both conceptual literacy and practical application, the book is organized into two interconnected parts. While each part offers distinct contributions, they are not discrete; thematic overlaps are intentional and reflect the interdependent nature of sustainability thinking. Part 1 lays the conceptual foundations for understanding sustainability in the THE sectors, while Part 2 builds on this with applied insights. The two parts are closely connected, with each informing the other, reflecting the integrated and iterative nature of sustainability learning. The chapters are not a fixed sequence of lessons, but invitations to revisit, reflect and apply ideas iteratively. This structure mirrors the complex, adaptive nature of sustainability work itself. Readers are encouraged to engage with sustainability not as a checklist, but as a value-driven dialogue. It also shows how slight discrepancies in data sets or statistics arise due to variations in context and sources, reflecting the nature of sustainability metrics and ongoing critical debate. Themes of justice, representation and human prospering are embedded throughout, offering a holistic view of what sustainability can and should mean in THE.

Part One: Framing sustainability for a changing industry

The first part of this book provides a foundation for understanding sustainability in the THE sectors, not as a theoretical abstraction or retrospective account, but as a way of interpreting and responding to the realities shaping the sector today. It invites readers to engage with sustainability as a professional mindset and transformative practice, one that shapes not only systems and policies, but also the values, behaviours and decisions that define leadership in THE. The chapters examine how sustainability is being defined, governed and debated within international contexts, and how professional competencies are evolving in response. Concepts such as systems thinking, cultural integrity and ethical responsibility are introduced as essential for navigating complexity. Rather than offering generalized principles, this section presents sustainability as a lived, professional concern that intersects policy, education and community engagement. Each chapter focuses on current relevance and application, enabling readers to connect global agendas with their future roles in a rapidly changing THE industry.

Chapter 2: Sustainability in the global context

This chapter introduces sustainability as a central concern for the THE sectors and explains why it has shifted from a marginal issue to a core strategic priority. It explores how environmental, social, cultural and economic pillars of sustainability intersect in THE practice and policymaking. Global frameworks and institutions, including the United Nations, UN Tourism and the European Commission, are examined for their influence on sustainability governance and regulation

The chapter positions sustainability within current global challenges such as climate change, inequality and responsible consumption, using systems thinking and critical reflection to interpret interdependencies. Readers apply the four pillars of sustainability to real-world examples and analyze how THE contributes to and is shaped by the Sustainable Development Goals, with particular focus on SDG 4 Quality Education and SDG 12 Responsible Production and Consumption. The role of competency frameworks in guiding professional development is introduced as a basis for navigating complex sustainability transitions. Readers are encouraged to reflect on global contexts not as abstract forces, but as living systems that shape the policies, pressures and possibilities that structure sustainability work within the sector today.

The real-world example explored in this chapter is sustainable tourism and destination recovery in Kaikōura, New Zealand.

Chapter 3: Professional development for sustainability

This chapter examines the role of professional development in preparing individuals to meet sustainability challenges in THE. It introduces key competencies such as critical thinking, systems thinking, ethical reasoning and strategic planning, highlighting the need for continuous development across a professional career. Tools such as the Learning Compass and training needs analysis are introduced to help readers identify current capabilities and design tailored development plans. Emphasis is placed on cultivating self-awareness and a values-driven mindset, positioning personal growth as integral to organizational sustainability.

The chapter explores how leadership grounded in sustainability principles can influence a culture of lifelong learning lifelong learning in a fast-evolving sector, where environmental pressures, regulatory expectations and stakeholder demands require adaptable and reflective practitioners. Professional development is not presented as a one-time exercise, but as a continuous process tied to strategic sustainability outcomes and long-term sector resilience. Readers are encouraged to view their growth as contributing not only to individual advancement, but to broader organizational and societal change.

The real-world example explored in this chapter is Green Pearls®, the communication and information platform for sustainable places for tourism.

Chapter 4: Responsible practice for cultural sustainability

This chapter explores the role of cultural heritage and identity within sustainability frameworks, focusing on how THE engages with culture as both a resource and responsibility. It examines how tangible and intangible heritage is shaped, protected or contested through tourism practices, and how poorly designed engagement can result in commodification, loss of authenticity and harm to local communities. Particular attention is given to the impacts of overtourism.

The chapter highlights the importance of ensuring that cultural voices are included in decision-making, planning and representation, with respect for local agency and lived experience. Readers explore how THE business practices influence social and cultural sustainability, and how corporate social responsibility and stakeholder theory provide models for ethical engagement. Cultural sustainability is also discussed in relation to the Sustainable Development Goals, reinforcing the role of THE in supporting inclusive, equitable and culturally rooted development. The chapter encourages a critical and reflective approach, equipping readers to consider both the risks and potential of cultural tourism as a vehicle for sustainable futures.

The real-world examples explored in this chapter are value creation at The IceHotel, Jukkasjärvi, Sweden, and overtourism in South Lake Tahoe, Sierra Nevada, USA.

Part Two: Applying sustainability in professional practice

The second part of the book turns its attention to the operational realities of sustainability in the THE sectors. While the first section provides the conceptual tools needed to understand sustainability, this section explores how those ideas are implemented, evaluated and embedded into routine business practice. The chapters examine how sustainability shapes decision-making, resource management and organizational strategy, with a focus on the daily contexts in which professionals operate. From food systems and circular design to performance reporting and ethical communication, the content highlights both the complexity and the necessity of applied sustainability. The aim is not only to inform, but to support the development of actionable knowledge and professional capability for effective, ethical and context-sensitive practice.

Chapter 5: Measuring sustainability and reporting

This chapter introduces the frameworks, tools and concepts used to assess and communicate sustainability performance in THE. It explores mechanisms such as corporate social responsibility and environmental, social and governance models as frameworks for defining ethical and sustainable practices. Readers examine how data is collected, analyzed and presented through sustainability reporting systems, and consider what counts as meaningful measurement in different operational contexts. The chapter introduces Scope 1, 2 and 3 emissions, auditing practices, and certification schemes, exploring their role in shaping transparency and accountability.

Readers are encouraged to critically evaluate how reporting frameworks influence perceptions of progress, and to reflect on the motivations behind different types of reporting. Measurement is not treated as a purely technical exercise, but as a process tied to decision-making and trust-building. The chapter prepares readers to engage with reporting both as future professionals who will implement it, and as reflective practitioners who must ask whether what is being measured aligns with sustainability goals and values.

The real-world examples explored in this chapter are the certification company EarthCheck and the Athenaeum Hotel and Residences, London, UK.

Chapter 6: The circular economy

This chapter explores the circular economy as a framework for rethinking production, consumption and waste in THE sectors. It contrasts circular principles with traditional linear models and introduces core concepts such as closed-loop systems, waste prevention and regenerative design. Readers are encouraged to apply systems thinking to identify how environmental, social and economic systems are interlinked, and to evaluate the consequences of maintaining resource-intensive practices.

The chapter introduces the 8Rs framework and explores how it supports net zero goals and innovation. Attention is also given to the role of collaboration as essential to enabling circular transitions across sectors, supply chains and stakeholders. Real-world examples illustrate how circular strategies are being applied in areas such as procurement, operations, and infrastructure. Readers are invited to reflect on how circular thinking can influence their own professional practice, and how it can be embedded into strategic planning and sustainability leadership within organizations and destinations.

The real-world examples explored in this chapter are the circular economy practices at luxury nature tourism Hotel Casa Palmela, Arrábida Natural Park, Portugal, and a Community Homestay Network operator in Nepal.

Chapter 7: Food waste

This chapter examines the global and sector-specific dimensions of food waste in THE, with a focus on its environmental, social and economic implications. It differentiates food loss and food waste and traces their occurrence along the supply chain – from sourcing and storage to preparation and service. Readers assess the drivers of food waste, including operational inefficiencies and human behaviours, and explore a range of interventions for prevention and reduction. These include portion control, inventory management, behavioural nudges and staff training.

The chapter also introduces sustainable approaches to managing unavoidable waste, such as composting and anaerobic digestion. Food waste is analyzed in connection with greenhouse gas emissions and the circular economy, positioning it as both a practical and moral sustainability issue. Readers are invited to consider how organizational culture, consumer expectations and systemic pressures contribute to waste, and how strategic, evidence-informed approaches can reduce waste while supporting financial, ethical and environmental outcomes.

The real-world example explored in this chapter is food waste reduction by cruise line operator P&O Cruises.

Chapter 8: Sustainable menus and ethical sourcing

This chapter equips the reader with the initial knowledge and understanding required to explore the development of sustainable menus in THE. Rather than presenting a single model for a sustainable menu, this chapter emphasizes an important learning point: sustainable menus must be tailored to the specific context and character of each organization. It encourages consideration of how sourcing decisions, preparation methods and menu design impact sustainability outcomes. Readers learn about the environmental and ethical implications of ingredients, including food miles, seasonality and traceability.

The chapter introduces practical approaches to integrating sustainability into menu planning, including the three-step framework and principles aligned with planetary-healthy diets. Operational realities such as cost, availability and consumer preferences are examined, alongside strategies to balance sustainability goals with business viability. Ethical sourcing is positioned as central to responsible menu development, connecting supplier relationships to broader sustainability values and standards. Through real-world examples and critical analysis, the chapter illustrates how menus can reflect not only culinary identity, but also a business's environmental and social commitments. Readers are encouraged to think strategically about how menus communicate values and influence behaviour, and how small design choices can contribute to broader sustainability narratives.

The real-world example explored in this chapter is the practical application of zero-waste principles at Michelin-star restaurant Silo.

Chapter 9: Greenwashing

This chapter investigates the practice of greenwashing and its implications for credibility, ethics and sustainability progress in THE. It explores how misleading or unsubstantiated environmental claims can undermine trust and obscure real sustainability performance. The chapter introduces the concept of greenhushing and discusses the tension between marketing ambition and regulatory compliance. Readers examine relevant legislation and guidelines designed to hold organizations accountable for their environmental claims and consider how businesses can communicate sustainability authentically. The ethical consequences of greenwashing are discussed in relation to stakeholder trust, consumer expectations and industry integrity.

Readers are encouraged to reflect on how sustainability narratives are constructed and challenged, and to identify strategies for transparent, accurate and meaningful communication. The chapter positions communication as a form of action that carries risk, responsibility and the potential to either reinforce or erode sustainability values.

The real-world example explored in this chapter is the UK's Advertising Standards Authority's ruling on airline company Ryanair.

Key features of this book

To help you get the most out of each chapter and connect your learning to the real world of THE, we have included several features throughout the book. Before starting the chapter, it is helpful to understand the key features included throughout. The features show how the contents link to real-world issues, and highlight the skills and knowledge needed for a career in THE. Each feature plays an important role in helping you to connect academic content with current industry expectations and practices. Each chapter follows a consistent structure to support your learning journey.

In addition to the main content, the book features a dedicated extension material chapter at the end. This chapter includes real-world examples and professional interviews that enrich and expand upon the themes explored throughout the book. Developed during the writing process, these resources are designed to provide further exposure to sustainability practices across diverse cultural, organizational and geographical contexts.

The extension material encourages reflection and deeper understanding. It can be used independently or alongside the main chapters, making it a valuable tool for self-guided learning.

What to expect in each chapter

Chapter aim

Each chapter begins with a clear aim that outlines the main focus and purpose. This helps guide your reading and gives context for the topics that follow.

Learning outcomes

Learning outcomes show what you should be able to do or understand by the end of the chapter. They support your ability to track progress and focus on key areas of learning.

Selected Sustainable Development Goals (SDGs)

Each chapter is aligned with Sustainable Development Goals most relevant to the content. The goals highlight how your studies connect to global priorities. They help you to understand the ethical and environmental responsibilities that come with working in the THE sector.

Key competencies for sustainability

These outline the skills and attributes that are developed through the chapter content. They reflect what is valued by employers in the THE sectors, such as communication, problem-solving, intercultural awareness and sustainability literacy.

Thinking

Being

Behavioural

Key words

Important terms and concepts are highlighted at the start or throughout the chapter. Understanding these key words helps you to build subject-specific vocabulary and improves your ability to engage with academic texts and professional discussions.

Industry voices

These boxes feature insights from professionals working in THE. They help connect theory to current practice and offer real-world perspectives from across the sector.

Real-world examples

Real-world examples are included throughout to demonstrate how theories and concepts apply in practice. These examples encourage you to think critically about current challenges, innovations and trends in the industry.

Activities

Activities are included to help you apply what you have learned, reflect on key ideas, and engage actively with the content. These tasks support the development of independent thinking and practical skills and are useful for both individual learning and group work.

Key takeaways

Each chapter ends with a summary of key takeaways. This section reinforces your understanding and highlights the most important ideas to remember.

Reflective questions, further resources and references

Finally, you will find reflective questions to engage and challenge your learning, further resources to explore and a list of references.

Student voice

Before beginning your own journey, first read a recent student's account of her work with sustainability in hospitality.

Meet Molly Eccleston, an undergraduate alumna of Edge Hotel School, at the University of Essex whose work exemplifies how student-led initiatives can address real-world challenges within professional THE settings. In her final year, Molly coordinated a team tasked with reviewing and enhancing the sustainability strategy of Wivenhoe House Hotel. The project brought together strategic analysis, stakeholder engagement and the principles of sustainable management, resulting in a strategy that was more transparent, measurable and inclusive. Molly's team offered a structured and practical approach to sustainability in hospitality operations at Wivenhoe House Hotel.

This work stands out as an example of excellent practice as it demonstrates how applied academic work can contribute meaningfully to an organization's development.

Readers of this textbook are encouraged to explore Molly's work below. It provides both insight and a grounded case of how hospitality students can drive meaningful change. As you read about her project, consider how your own learning might evolve through similar opportunities, and let this example guide you through the rest of the book.

Molly Eccleston is a Head Start apprentice at Hallgarten and Novum Wines in the UK.

REAL-WORLD EXAMPLE   Sustainability project at Wivenhoe House Hotel

Wivenhoe House Hotel is home to Edge Hotel School, offering hospitality and events management degree courses at the University of Essex. This unique concept allows undergraduate students to undergo work placements in all departments, offering a holistic perspective of the hotel. Alongside class projects that encourage critical evaluation of how the hotel operates, this enables Edge students to confidently pursue supervisory and management opportunities in THE industry on completing their degree. Students are integral in making sustainability improvements to the hotel by applying theory learnt in class. Some ways in which they have left their mark include replacing plastic room cards with ones made from wood, recycling coffee grounds and introducing a small kitchen garden.

My final year hotel management class was tasked with enhancing the sustainability strategy of Wivenhoe House. Over three weeks, we reviewed and restructured the strategy with the aim of transforming the hotel into a sustainability asset for the University of Essex.

I took on the role of co-ordinator and delegated different aspects of the project between team members. One person took charge of analysing Wivenhoe House's local competitors and their approaches to sustainability. They focused on three hotels, all with four-star ratings, offering a similar customer experience to the same target market. This research found a lack of comprehensive sustainability strategies and accreditations amongst competitors, which highlighted the opportunity for Wivenhoe House to differentiate themselves using their sustainability strategy. Whilst they have an existing policy, some local awards and a relatively high sustainability rating on Booking.com, there were still areas that Wivenhoe House could improve further. This was the focus of the other members of the team.

One person took responsibility for evaluating the effectiveness of Wivenhoe House's existing strategy and comparing this to the University of Essex's sustainability sub-strategy (2021–26). Issues with the current strategy included a lack of statistics, graphs and other evidence, meaning claims appeared vague and unsubstantiated. The use of only future tense also made it difficult to differentiate between actions which Wivenhoe House had already taken and initiatives which they aimed to pursue in the future. Overall, the strategy focused on breadth, rather than depth. This was in direct contrast to the University of Essex sub-strategy, which focuses on 13 priorities, supporting each with tangible key performance indicators, resulting in a transparent and trustworthy document.

To align Wivenhoe House's strategy with that of the University, we renamed the document 'Wivenhoe House Hotel Sustainability Strategy (2021–2026)'. Vision, mission and value statements were then established, to provide clear direction for the strategy. Our group suggested that these statements could be contributed by staff members, to reflect their aspirations for the hotel, the realities of their day-to-day work and how they feel the business should operate. By giving staff an active role in creating the strategy, they are more likely to take responsibility for delivering it.

The inspiration for the restructuring of the strategy was a Triple Bottom Line approach. Widely adopted within corporate strategies, Triple Bottom Line places equal importance on Profit, Planet and People. This is important for Wivenhoe House as their long-term survival is not only dependent on financial success, but also on creating value for the university, students and the community, all important stakeholders.

We decided to enhance the Triple Bottom Line by adding Plate, focusing on all food and drink sustainability issues. These four aspects became the basis of the strategy and each was aligned with one of the university's 13 sustainability priorities. For Profit, we focused on waste management, for planet, reducing Scope 1 and 2 emissions in line with the UN's sustainable development goals, whilst Plate emphasized responsible sourcing, preparing and serving. The People section prioritized education for sustainability, demonstrating the hotel's responsibility to their students, as a partner of the university. By educating students about the importance of sustainability in THE industry, Wivenhoe House will play an active role in shaping the way they operate as future managers.

Prioritizing these four issues ensures that the strategy is more in-depth and provides a framework for decision making and resource allocation. Each priority was also split into past and future actions, to increase transparency, and measured using key performance indicators. Providing evidence of Wivenhoe House's progress, quantifiable targets will show that the strategy is not just a box-ticking exercise, gaining trust from stakeholders. Examples of these include helping the university to achieve a top 30 position in the People and Planet University League by 2027 and reduce CO_2 emissions by 45 per cent by 2030.

The final member of the group focused on recommending future initiatives for the hotel. This included appointing a 'green ambassador' for each department to champion sustainability. These ambassadors could encourage other members of their department to engage with the strategy and feed back to a committee with ideas to make the hotel even more sustainable.

Our findings and recommendations were presented to senior management, with the aim of building buy in from those integral in decision making. The presentation focused on communicating the role of the strategy in formalizing Wivenhoe House's commitment to balanced stakeholder management, creating value for all stakeholders, including staff, students, the local community and their environment. It also demonstrated the benefits for management, as becoming a sustainability asset could strengthen their partnership with the university. Our presentation was well received and moving forward we will work with senior management to action our suggested changes.

This task was also beneficial for us as management students. Our research highlighted the many ways in which a hotel can impact its environment, including excessive consumption of resources and production of waste, and the subsequent importance of comprehensive sustainability strategies within the THE industry. Providing knowledge and skills that can be carried forward into our future management careers, this task not only reflects the 'learn by doing' ethos of Edge Hotel School, but also exemplifies how sustainability best practice can be implemented within hospitality higher education. The opportunity to make a real, positive difference ensures that sustainability will be prioritized by students in the future.

References

Ellen MacArthur Foundation (2021) Completing the picture: How the circular economy tackles climate change, 26 May, https://ellenmacarthurfoundation.org/completing-the-picture (archived at https://perma.cc/QSR2-SV9L)

Gabbitas, A (2023) Staff retention and training in hospitality, *Hospitality Workforce Review*, **15** (2), pp 45–52

Oxford Economics (.n.d.) The environmental and social impact of travel and tourism: Methodological report for the World Travel & Tourism Council, https://globaltravelfootprint.wttc.org/ (archived at https://perma.cc/3XR6-Z3T5)

UN (1992) United Nations Conference on Environment & Development: Agenda 21, 3–14 June, https://sustainabledevelopment.un.org/content/documents/Agenda21.pdf (archived at https://perma.cc/AU9W-5MEX)

UN (2015) Transforming Our World: The 2030 Agenda for Sustainable Development, https://docs.un.org/en/A/RES/70/1 (archived at https://perma.cc/ESG9-A438)

UNEP (2024) Emissions Gap Report 2024: No more hot air…please! www.unep.org/resources/emissions-gap-report-2024 (archived at https://perma.cc/4H97-Y7ZA)

UN Tourism (.n.d.) www.unwto.org/ (archived at https://perma.cc/Z9L7-LLZ9)

World Commission on Environment and Development (1987) *Our Common Future*, Oxford University Press, Oxford

WTTC (2024) Economic Impact Research 2024: Global Trends, 4 September, https://researchhub.wttc.org/product/economic-impact-report-global-trends (archived at https://perma.cc/6JZS-WUUW)

WTTC (.n.d.) Travel & Tourism Economic Impact Research (EIR), https://wttc.org/research/economic-impact (archived at https://perma.cc/4W8H-LJR8)

Yu, M, Li, X, Huang, D and Zhou, Y (2025) Global food transport and tourism-related emissions: An empirical assessment, *Journal of Sustainable Consumption*, **11** (1), pp 67–84

PART ONE
Framing sustainability for a changing industry

2 | Sustainability in the global context

| Systems thinking competency | Anticipatory competency (Future thinking) | Critical thinking competency | Collaboration competency |

CHAPTER AIM

The aim of this chapter is to examine the evolution of sustainability within the global tourism, hospitality and events (THE) sectors, as shaped by international policy and governance. It traces key historical and contemporary developments led by the United Nations (UN), the UN World Tourism Organization (now UN Tourism), and the European Commission which have positioned sustainability as a core principle in THE. Through these global initiatives, the economic, environmental, social and cultural dimensions of THE have become increasingly recognized as integral to sustainable development. THE has been explicitly identified as a catalyst for achieving the Sustainable Development Goals (SDGs). This chapter is underpinned by SDG 4 Quality Education and SDG 17 Partnerships for the Goals as it contributes to learner knowledge and understanding of the impact and responsibilities of THE for sustainable development.

The chapter aims to support the learner to develop the following four key competencies for sustainability:

- Systems thinking competency – to enable the identification and interpretation of the relationships within the global context in which THE operates.
- Anticipatory (future) thinking competency – to understand and consider future implications of THE practices.
- Critical thinking competency – to encourage the evaluation of past and present sustainability efforts so that a point of view can be taken about the roles and responsibilities of THE.
- Collaboration competency – to learn from others and recognize the importance of engaging with diverse stakeholders in sustainable development processes.

LEARNING OUTCOMES

Upon completion of this chapter, you will be able to:

- explain the historical development of sustainability in the global context of tourism, hospitality and events
- demonstrate critical thinking competency by identifying and evaluating the influence of international institutions such as the UN, UN Tourism and the European Commission on sustainability policies and practices in THE sector
- apply the four pillars of sustainability (economic, environmental, social and cultural) to a real-world THE example using systems thinking and critical thinking competencies
- analyze how THE contributes to and is impacted by the Sustainable Development Goals, with particular focus on SDG 4 Quality Education and SDG 12 Responsible Consumption and Production
- anticipate sustainable pathways for the sector by interpreting and using sustainability competency frameworks
- collaborate effectively to assess and propose sustainable actions within THE scenarios

KEY WORDS

Sustainable THE, THE governance, international frameworks, policy development, cultural heritage, sustainable development goals, competence frameworks, next tourism generation.

Introduction

This chapter explores the evolution of sustainability within the global context of the tourism, hospitality and events (THE) sectors, following a milestone timeline that begins with the 1972 Stockholm Conference and concludes with the 2024 introduction of the Statistical Framework for Measuring the Sustainability of Tourism. It focuses on the influence of key international frameworks, policies and governance mechanisms. It traces how sustainability has become central to global development agendas and how, in turn, this has shaped practice within THE.

The timeline presents key moments that have defined the sustainability discourse. These include developments such as the establishment of the United Nations Environment Programme (UNEP) and the United Nations World Tourism Organization (UNWTO), influential publications like the Brundtland Report and Agenda 21, and key models such as the Triple Bottom Line.

From 2001 to 2005, UNESCO-led work on culture and heritage framed culture as a critical dimension of sustainable development. The declarations and conventions are presented together to reflect conceptual continuity and the shared objectives across these instruments, which collectively frame culture as both a foundation and facilitator of sustainable development. It also enables a more cohesive exploration of their practical relevance to THE sectors.

More recent developments underscore the increasing integration of sustainability into policy implementation and professional practice. For instance, the European Union's Transition Pathway for Tourism and the subsequent Sustainable EU Tourism project are discussed jointly to illustrate how strategic frameworks can evolve into applied initiatives. Instruments such as GreenComp and the Pact for Next Tourism Generation Skills show a shift toward defining and supporting sustainability competencies across sectors.

Table 2.2 represents the developments as a connected timeline illustrating how sustainability in THE has progressed from abstract principle to operational imperative, shaped by global commitments, regional initiatives and sector-specific strategies.

1972: Stockholm Conference and United Nations Environment Programme (UNEP)

The 1972 United Nations Conference on the Human Environment, held in Stockholm, marked a turning point in global environmental governance. As the first international dialogue between industrialized and developing nations focused on addressing environmental issues, the conference produced the Stockholm Declaration with 26 guiding principles and an action plan. The action plan was organized under three main categories: Environmental Assessment, Environmental Management and International Supporting Measures (UN, .n.d. a).

A direct outcome of the Stockholm Conference was the creation of the United Nations Environment Programme (UNEP). UNEP shaped global environmental policies to assist countries in their implementation. The polices focused on climate action, nature conservation, pollution reduction and green economic development. Its remit directly addressed the Stockholm Conference's action plan on environmental assessment, environmental management and governance, and education and training.

The Stockholm Conference and UNEP laid the ethical and strategic groundwork for environmentally responsible practices in THE. Whether operating locally or internationally. THE businesses are stakeholders in maintaining environmental and human stability. Understanding UNEP's priorities helps industry professionals recognize the importance of incorporating sustainability in operations, decision-making and long-term planning.

1975: The UN World Tourism Organization (UNWTO)

Formerly the International Union of Official Travel Organizations (IUOTO), the UN World Tourism Organization (UNWTO) was established in 1975. UNWTO became the leading international body for global tourism governance, with a mandate to promote responsible, sustainable and universally accessible tourism. UNWTO positioned tourism as a driver of economic growth, inclusive development and environmental sustainability especially in developing regions (UN Tourism, .n.d.). In 2003 it became a specialized agency of the UN, strengthening its influence and commitment to global development agendas.

Rebranded as UN Tourism in 2024, the organization continues to support tourism through policy development, capacity-building and knowledge sharing. Its initiatives include the Tourism Satellite Account, a standard statistical framework (UNWTO, 2008), and the Statistical Framework for Measuring the Sustainability of Tourism (MST) (UN Tourism, 2024), the main tools for the economic measurement of tourism.

UN Tourism set the strategic direction for sustainable tourism practices. For THE professionals, understanding its work is crucial for career proficiency. It provides the frameworks and tools for aligning tourism operations with global sustainability goals. It shapes how the THE sector responds to climate change, social equity and cultural preservation.

1987: The Brundtland Report: *Our Common Future*

Published in 1987, the Brundtland Report – officially titled *Our Common Future* – was key to framing sustainability as a global imperative. Described as 'a global agenda for change' (World Commission on Environment and Development, 1987), it called upon political and economic institutions in the public and private sectors, at global, national, and local level to take unified action toward sustainable development. One of the most significant contributions of the Brundtland Report was its definition of sustainable development as 'development that meets the needs of the present without compromising the ability of future generations to meet their own needs' (World Commission on Environment and Development, 1987). Although this definition has been debated for its ambiguity, it remains a cornerstone of sustainability discourse.

The report emphasized the interconnectedness of environmental health, social justice and economic development – an approach recognized as the three pillars of sustainability. For THE sectors this means balancing profitability with environmental responsibility and equitable social outcomes.

The Brundtland Report is core for THE professionals aiming to embed sustainability into operations. Its principles promote long-term thinking, integrated policy approaches and global cooperation. These ideas continue to influence sustainability standards such as corporate social responsibility (CSR) and environment, social and governance (ESG) models of business practice.

1989: The Hague Declaration on Tourism

Adopted by UNWTO in 1989, the Hague Declaration marked the first international policy specifically focused on sustainable tourism. Framing tourism as an opportunity and a responsibility for sustainability, the Declaration laid out ten guiding principles.

Notably, Principle III is directly linked to environmental impact of THE as it states 'An unspoilt natural, cultural and human environment is a fundamental condition for the development of tourism. Moreover, rational management of tourism may contribute significantly to the protection and development of the physical environment and the cultural heritage, as well as to improving the quality of life.' (UNWTO, 1989). This principle expresses the duality of tourism – it depends on but can also help protect environmental and cultural resources.

The Hague Declaration remains applicable to THE sectors because it emphasizes international cooperation, active community involvement, environmental protection and continuous monitoring of impact. For THE professionals, it encourages engaging local stakeholders in making evidence-based decisions. It highlights the responsibility of THE for environmental stewardship.

1992: Earth Summit and Agenda 21

Agenda 21, described as 'a dynamic programme' (UN, 1992), emerged from the United Nations Conference on Environment and Development (UNCED), also known as the Earth Summit, in 1992. It represented a global consensus and political commitment to rethink economic growth, advance social equity and ensure environmental protection by promoting global sustainable development.

Building on the Brundtland Report, Agenda 21 stressed the need for integrated policy action at global, national and local levels. It promoted institutional and governance reform and emphasized that environmental, social and economic challenges must be addressed holistically.

Agenda 21 remains a central document for THE because it advocates:

- the involvement of Indigenous people in managing and planning local environments, respecting traditional knowledge and customs
- the promotion of 'environmentally sound and culturally sensitive tourism programmes as strategies for sustainable development' (UN, 1992)
- collaboration between governments, businesses and industries to use economic instruments and market mechanisms to guide sustainable tourism

Agenda 21 also warns against the degradation of fragile ecosystems – forests, mountains, coastal and marine environments – highlighting the importance of tourism management in preserving biodiverse ecosystems. Objective 13.15 clearly states that sustainable tourism must also ensure the protection of the livelihoods of local communities and Indigenous peoples.

For THE professionals, Agenda 21 calls for eco-tourism, the protection of local communities and the integration of sustainability in tourism planning. It links directly to THE by emphasizing education, public awareness and the application of sound environmental policies, aligning with the Hague Declaration and the ongoing work of UNEP and UN Tourism.

1994: Elkington's Triple Bottom Line (TBL)

In 1994 John Elkington, the founder of SustainAbility, introduced the concept of the 'Triple Bottom Line' (TBL), challenging business to move beyond profit-only thinking.

Also known as the Triple-P model, TBL asserted that organizations must balance Profit, People, and Planet for positive change. Elkington argued that only when these three dimensions align can long-term sustainability be realized (Elkington, 2004).

For Elkington, Profit remained essential, but he argued it must be achieved through strategies that also drive social and environmental responsibility. Economic viability depended on sustainable practices that drove innovation and efficiency. The People dimension focused on a business's societal impact, emphasizing stakeholder, not just shareholder, value. Strategic partnerships with communities and non-profits should result in equitable benefits. The third dimension was concerned with the positive impact business could have on the Planet by reducing their environmental footprints. For THE sectors, this means adopting eco-friendly operations, ethical sourcing and committing to climate change.

The TBL remains a core framework for THE. It informs certification schemes, sustainability reporting and operational strategies. While some critique it as idealistic in a profit-driven world, TBL provides THE professionals with a powerful tool to navigate stakeholder expectations, reduce impacts and create future-ready businesses.

ACTIVITY 1: EXPLORING THE STAKEHOLDERS AND THE TRIPLE BOTTOM LINE

Below is a list of key stakeholders in the THE industries.

Instructions

1 **For each stakeholder**, identify which dimensions of the TBL (Profit, People, Planet) they most influence.

2 **Briefly explain how they contribute** to sustainable THE practices.

3 Try to think critically:

   o Profit: How do they contribute to or benefit from economic success?

   o People: How do they impact or represent social values, equity and wellbeing?

   o Planet: What role do they play in protecting or harming the environment?

Stakeholder	Dimensions of TBL in order of priority	How they relate to TBL
Government agencies	*People* *Planet*	*Develop policies, infrastructure and regulations supporting sustainable development.* *Responsible for community consultation and environmental stewardship.*

(continued)

(Continued)

Stakeholder	Dimensions of TBL in order of priority	How they relate to TBL
Local communities	Profit	Impacted directly by tourism
	People	Bring essential local knowledgeAdvocate for fair benefit-sharing and environmental preservation.
	Planet	
Accommodation providers		
Attractions and entertainment venues		
Environmental and cultural preservation organizations		
Restaurants and food service providers		
Retailers and souvenir shops		
Tourism industry associations, boards and authorities		
Tourism research and consultancy organizations		
Tour operators and travel agencies		
Tourists / travellers / visitors		
Transportation providers (airlines, coaches, trains)		

2001–05: Cultural diversity and the role of Tourism, Hospitality and Events (THE) in sustainable development

Across international sustainability frameworks, culture has increasingly been recognized not only as heritage to be preserved but as a dynamic and living resource critical to sustainable development. Within THE sectors, promotion, protection and respect for cultural diversity are fundamental responsibilities. A coherent thread runs through key UNESCO documents – the 2001 Universal Declaration on Cultural Diversity, the 2003 Convention for the Safeguarding of Intangible Cultural Heritage and the 2005 Convention on the Protection and Promotion of the Diversity of Cultural Expressions – that collectively set out principles essential for embedding cultural sustainability into THE practices.

The Universal Declaration on Cultural Diversity (2001) was a landmark document, establishing cultural diversity as a 'common heritage of humanity', The Declaration defined culture as 'the set of distinctive spiritual, material, intellectual and emotional features of society or a social group, and that it encompasses, in addition to art and literature, lifestyles, ways of living together, value systems, traditions and beliefs' (UNESCO, 2023). This holistic understanding of culture moves beyond tangible artifacts to include living expressions, community values and evolving traditions. It aligns closely with THE's objective to create meaningful visitor experiences rooted in the authentic life of host communities.

Building on the foundations of the 2001 Declaration, the 2003 Convention for the Safeguarding of the Intangible Cultural Heritage (ICH) placed emphasis on the dynamic nature of ICH which included practices, oral traditions, ritual, festive events, crafts, and knowledge related to the country. The Convention provided a legal framework to safeguard ICH, preserving cultural diversity by protecting the past and ensuring living traditions thrived and adapted for the future. The 2003 Convention highlighted that communities themselves must identify and safeguard their heritage. It acknowledged that safeguarding measures should be community-driven, participatory and respectful of the context in which heritage is practiced.

The 2005 Convention on the Protection and Promotion of the Diversity of Cultural Expressions further advanced these ideas by framing cultural diversity as integral to human development. It moved beyond safeguarding to actively promoting environments where cultural industries and expressions could flourish equitably. The Convention laid out the importance of culture for sustainability as 'recognizing the importance of traditional knowledge as a source of intangible and material wealth, and in particular the knowledge systems of indigenous peoples, and its positive contribution to sustainable development, as well as the need for its adequate protection and promotion'(UNESCO, 2005). The 2005 Convention positioned culture as both a driver and enabler of sustainable development, requiring proactive engagement not just from governments but also from civil society and private sector actors, including those within the THE industries.

Together, the declaration and conventions form a cohesive framework for understanding cultural sustainability, and the 2003 and 2005 Conventions can be seen as deepening and operationalizing the vision first laid out in 2001. They affirm that safeguarding cultural diversity is not an optional add-on for THE, but a fundamental responsibility linked to broader goals of human rights, community empowerment and sustainable economic growth.

For THE professionals, there are implications for business practice. First, there is a need to view cultural expressions not as commodities to be consumed by tourists, but as dynamic systems of knowledge, creativity and identity. Tourism businesses, for example, have the power to either commodify or authentically showcase local traditions. Second, THE initiatives must actively support local cultural economies. Whether through sourcing locally produced handicrafts, employing local artists and performers or collaborating with Indigenous knowledge holders, THE operations can contribute to the vitality of cultural expressions. An illustrative example is found in New Zealand, where Māori protocols are integrated into formal events and ceremonies. The use of Māori greetings – *Tēnā koe* for one person, *Tēnā kōrua* for two people, and *Tēnā koutou* for groups – demonstrates how linguistic respect becomes a gateway for deeper cultural understanding and appreciation (Moorfield, 2011).

Finally, these frameworks cautioned against the dangers of cultural homogenization and loss. In an increasingly globalized world, mass tourism can erode distinctive cultural identities unless carefully managed. Overall, the Universal Declaration on Cultural Diversity (2001) and the subsequent Conventions of 2003 and 2005 collectively shaped an integrated vision of cultural sustainability that continues to be directly relevant to THE.

2010: The pillars of sustainability

The concept for the four pillars of sustainability provided a structured framework to guide individuals, businesses and governments towards practices that built on the foundations laid by the Brundtland Report (1987) to ensure the wellbeing of both present and future generations.

The four pillars model evolved from Elkington's Triple Bottom Line model, which encouraged business to move beyond purely financial reporting and assess their social and environmental impacts alongside profits. Later UNESCO's Universal Declaration on Cultural Diversity (2001) introduced culture as the fourth critical dimension, recognizing that cultural identity and diversity are essentials components of sustainability.

The four pillars – environmental, social, cultural and economic – are widely used across business sectors, including THE. They serve as flexible yet interconnected principles that organizations adapt to their contexts. Importantly, the pillars are not independent; if one pillar is weakened, the others come under stress. If one collapses, the pursuit of a sustainable future is compromised. Figure 2.1 illustrates the co-reliance of the four pillars. This fragility is evident with the declaration of a global climate emergency highlighting the vulnerability of the environmental pillar.

Figure 2.1 The four pillars of sustainability

In the context of THE, the four pillars emphasize that environmental stewardship, social wellbeing, cultural preservation and economic growth must be achieved simultaneously. This is particularly relevant to THE where destinations, communities, and ecosystems are integral to the tourism product itself. Without prioritizing environmental sustainability, for example, the social, cultural and economic pillars that depend upon it cannot be maintained. This is especially significant for THE operating both domestically and internationally.

Pillar of environmental sustainability

Environmental sustainability involves implementing sustainable practices to minimize resource consumption, conserve biodiversity and preserve natural habitats through eradicating carbon emissions and other types of toxic pollutants. For THE, this is especially critical given its reliance on natural resources and ecosystems.

Some examples of best practice emerging in THE include:

- implementing eco-friendly transportation options for guests
- promoting energy-efficient technologies in accommodation
- reducing water consumption and introducing recycling and composting initiatives at events

Environmental stewardship in THE has to extend to include supporting local conservation research. For instance, promoting attractions that reinvest profits into protecting local wildlife areas would not only enhance sustainability but also improve the visitor experience. Natural capital, such as clean air, pollution free oceans and balanced biodiversity, is the foundation upon which THE destinations operate. If degraded, THE as an industry risks losing its core assets.

Pillar of social sustainability

Social sustainability focuses on building inclusive, equitable communities and protecting human health, education and human rights. In the THE context, this means ensuring that business practice benefits host communities, employees and guests alike.

From a business good practice perspective this would include:

- Providing fair wages, just contracts and safe working conditions. This is particularly important in an industry dominated by short-term and fixed term employment.
- Promoting employee health and wellbeing to enhance job satisfaction and reduce employee turnover.
- Engaging with local communities to ensure THE development enables social cohesion, protects community identity and shares benefits equitably.

Social sustainability also connects to the environment. Introducing healthier dietary choices on menus, for example, contributes to better public health while reducing the ecological footprint associated with food production, consumption and waste.

Pillar of cultural sustainability

Cultural sustainability emphasizes the importance of nurturing and supporting cultural heritage, traditions and diversity. In this way, THE businesses need to consider themselves as a member of society. The pillar requires THE to understand and respect the cultural identities of the host country in which they operate. In terms of appropriate operational practice this would consider:

- engaging meaningfully with host communities to understand and promote authentic cultural experiences
- supporting the preservation of traditional knowledge, sites and customs through tourism revenue and partnerships
- encouraging cultural sensitivity among staff and guests, including language skills development, which research (Žerajić, 2020) shows improves service quality and reduces cultural misunderstandings

Recognizing the integral nature of culture for both community identity and unique tourism experiences enhances both social cohesion and economic development for THE.

Pillar of economic sustainability

By understanding the importance of focusing on the environmental, social and cultural pillars, THE will facilitate creating systems that support long-term economic health. Economic sustainability ensures that THE operations are profitable but also financially benefit the broader community and environment. It includes considerations of conditions of employment, innovation and creativity, fair trade, and responsible production.

Examples of economically sustainable practices include:

- hiring and training local people across all levels of the organization
- sourcing goods and services from local enterprises to strengthen community economies
- promoting local cultural and environmental assets while ensuring minimal negative impact

Economic sustainability, when built on environmental, social and cultural pillars, creates opportunities for long-term viability and competitiveness. Drawing upon local knowledge, economic sustainability in THE can be strengthened by the co-creation of unique, niche offerings such as quiet escapes, noctourism (night-time travel experiences) and culinary experiences. Differentiating destinations enables the targeting of high value markets (Hall, 2025).

In THE, using the framework of the four pillars can offer a comprehensive assessment of host communities environmental, societal, cultural and economic dimensions. It can facilitate the formulation of clear sustainability objectives and introduce diverse points of view into sustainability initiatives that benefit both the communities and the business.

REAL-WORLD EXAMPLE Kaikōura, New Zealand

Background: Kaikōura, located on the east coast of New Zealand's South Island, is a small coastal town internationally renowned for marine-based tourism, particularly whale watching.

In 2016, Kaikōura experienced a devastating earthquake considered to be the 'most complex quake ever studied' (Amos, 2017), which severely damaged infrastructure, left Kaikōura and its neighbouring communities cut off, and disrupted the local economy (Institute of Civil Engineers, .n.d.). With a resident population of just over 2,000 people, Kaikōura's economy was reliant on tourism, fishing and farming (Statistics New Zealand,

2018). Kaikōura's resident community and local iwi (Indigenous community) committed to rebuilding sustainably. The result is considered a leading example of how destinations can successfully integrate the four pillars of sustainability into a resilient and thriving tourism model.

Pillar of environmental sustainability

In 2002, Kaikōura became the first local authority in New Zealand to achieve EarthCheck Certification, an international benchmark for sustainable tourism practices. Since then, the community has implemented rigorous waste management systems, marine ecosystem protection initiatives, and sustainable resource use strategies (Eco-Business, 2014).

A key contributor to Kaikōura's environmental leadership is Whale Watch Kaikōura, a tourism business owned and operated by the Indigenous Ngōti Kuri iwi. Whale Watch has gained international recognition for its commitment to conservation and for reinvesting tourism revenue into marine research and ecosystem stewardship. Its purpose-built vessels serve both tourists and scientists, collecting real-time data on whale populations, migration patterns and marine health. The information is relayed to the tourists on board in real time. In this way, tourists are aware that the money they pay for whale watching directly supports the continued research and conservation of marine life (Whale Watch Kaikōura, .n.d.). This model not only enhances visitor education but also funds local environmental initiatives and contributes updated research to New Zealand's Department of Conservation (Tarantino, 2023). Through these efforts, Kaikōura demonstrates how tourism can be integrated with environmental protection to ensure long-term ecological sustainability.

Pillar of social sustainability

In the aftermath of the 2016 Kaikōura earthquake, the community's social sustainability efforts were pivotal in building resilience and recovery. Local initiatives focused on community wellbeing, empowerment and inclusive participation in the rebuilding process.

One example is the establishment of the Kaikōura Social Recovery Plan, developed by the Kaikōura District Council. This plan aimed to empower individuals and communities to adapt positively to their changing environment. It encompassed various aspects, including psychosocial support, accommodation, financial assistance, community participation, communication and engagement. The plan emphasized the importance of community involvement in decision-making processes, ensuring that recovery efforts aligned with the needs and values of the residents.

Additionally, the Takahanga Marae (building complex) played a crucial role as a Civil Defence welfare centre. Immediately after the earthquake, the marae provided shelter, food and support to hundreds of people, demonstrating the strength of community networks and the significance of culturally appropriate responses in disaster situations (Thusoo et al, 2019).

Pillar of cultural sustainability

Kaikōura's approach to cultural sustainability is rooted in the recognition of its Indigenous Maori heritage. The town's Destination Management Plan (DMP) emphasizes the importance of integrating cultural values into tourism development, ensuring the identities and traditions of the local iwi are preserved and respected (RTNZ, 2025).

One initiative was the development of astrotourism: 'travel that takes people to somewhat remote locations … in their search for unpolluted views of the cosmos' (Night Sky Tourist, 2019). The Kaikōura Dark Sky Sanctuary was accredited by the International Dark Sky Association as a Dark Sky destination in 2024. This sanctuary not only promotes astrotourism but also aligns with Māori principles of 'kaitiakitanga' (guardianship). The sanctuary's creation involved community engagement and collaboration with local iwi, highlighting the integration of cultural values in environmental conservation efforts.

Additionally, the town has implemented cultural artwork installations along its coastal areas, such as pouwhenua (carved wooden posts) and interpretive panels that share Māori legends and local history. These installations serve both as educational tools for visitors and as a means of honouring and perpetuating the region's cultural narratives.

Pillar of economic sustainability

The 2016 Kaikōura earthquake caused extensive disruption to the region's economy, particularly affecting tourism, hospitality and small businesses. In response, Kaikōura prioritized rebuilding critical infrastructure through the North Canterbury Transport Infrastructure Recovery (NCTIR) alliance, which restored State Highway 1 and the Main North Rail Line – essential for reconnecting the town to visitors and supply chains.

Rather than simply reconstruct what was lost, Kaikōura embraced the 'Build Back Better' framework, aiming not just to restore but to improve its economic foundations (Neeraj, Mannakkara and Wilkinson, 2021). The framework encouraged community involvement in planning decisions, promoting sustainable development principles that would enhance resilience to future shocks. Support for small businesses was central to recovery efforts. Grants, advisory services and employability initiatives helped local enterprises adapt and ensured economic benefits remained within the community.

This future-oriented approach was evident in the 'Reimagine Kaikōura' plan, a strategic initiative designed to revitalize the tourism sector through diversification and sustainability. The plan supported local businesses to innovate, encouraged eco-tourism ventures, and invested in community-led projects that integrated environmental conservation with economic opportunity.

Kaikōura's recovery after the 2016 earthquake demonstrates that sustainable rebuilding requires balancing all four pillars of sustainability. By restoring the environment, strengthening community resilience, protecting cultural identity and ensuring local economic revitalization, Kaikōura created a more sustainable, resilient destination prepared for future challenges.

> **REFLECTIVE QUESTIONS**
>
> Based on the Kaikōura real-world example, reflect on the following questions to deepen your understanding of sustainable tourism and destination recovery. Respond in short paragraphs using specific information from the example:
>
> 1 **Environmental sustainability**: How can integrating environmental conservation into disaster recovery plans enhance long-term ecological resilience for tourism destinations?
> 2 **Social sustainability**: What strategies can communities employ to maintain social cohesion and support vulnerable groups during and after disaster recovery?
> 3 **Cultural sustainability**: How can incorporating Indigenous art, language and storytelling into public spaces contribute to cultural preservation and sustainable tourism?
> 4 **Economic sustainability**: How can disaster-affected communities balance immediate economic recovery needs with long-term sustainability goals to build greater resilience against future crises?

> **INDUSTRY VOICE: WALTER LEAL FILHO**
>
> Walter Leal Filho, Professor, HAW Hamburg and Manchester Metropolitan University, says: 'Managing the complexity of SDG implementation is not simple. As they are structured, their implementation depends on partnerships and engagements of local communities and a wide range of stakeholders and require proper sustainability leadership.'

2015: The 2030 Agenda for Sustainable Development and the Sustainable Development Goals (SDGs)

The 2030 Agenda for Sustainable Development was a global action plan adopted by all 193 United Nations member states in 2015. It was developed in response to the urgent need for a shared, comprehensive approach to global challenges such as poverty, inequality, environmental degradation and climate change. The 2030 Agenda aimed to be ambitious and practical by offering a vision of sustainable development that was a 'win-win cooperation which can bring huge gains to all countries and all parts of the world' (UN, 2015).

Central to the 2030 Agenda were the Sustainable Development Goals (SDGs), a set of 17 interconnected global goals which provided a shared language for governments, businesses and communities to address global challenges while aiming 'that no one will be left behind' (UN, 2015). While initially launched in 2015, the relevance and application of the goals continues to be used to shape current global efforts. The goals provide a structured framework to guide national and international efforts through to 2030.

The SDGs are supported by 169 specific targets and over 230 indicators that enable governments, industries and civil society to plan, implement and monitor their progress. They reflect recognition that sustainability cannot be achieved by addressing the environmental, social (including cultural) and economic pillars in isolation, but must instead involve an integrated approach that balances all three.

A cohesive framework for sustainable development

The 2030 Agenda is a complex document in the way its components, including the breadth of the SDGs, are designed to function as an integrated whole. Three components are central to this cohesion:

1 The four core principles providing the foundations for the implementation of the 2030 Agenda.
2 The Five Ps functioning to structure the 2030 Agenda's priorities.
3 The SDGs sitting as the operational core.

The four core principles: Foundations for implementation

The 2030 Agenda is underpinned by four guiding principles (UN, .n.d. b) that determine how the SDGs should be interpreted and implemented:

- **Integration**: Recognizing that challenges such as poverty, climate change and inequality are interdependent and must be addressed holistically.
- **Universality**: Affirming that the 2030 Agenda applies to all countries, regardless of income level, and requires action for every sector of society.
- **Inclusivity**: Ensuring that all people, especially those most vulnerable and marginalized, are included in development processes.
- **Equity**: Promoting fairness, social justice and equal access to resources and opportunities.

These principles provide an ethical and strategic foundation for sustainable development. They ensure that implementation of the 2030 Agenda is fair, participatory and responsive to local and global contexts.

The Five Ps: Structuring priorities

To make the 2030 Agenda accessible and actionable, the 17 SDGs are grouped into five overarching themes, often referred to as the Five Ps:

- **People**: Ending poverty and hunger, and ensuring health, education, dignity and equality for all:
  - SDG 1: No Poverty
  - SDG 2: Zero Hunger
  - SDG 3: Good Health and Well-being
  - SDG 4: Quality Education
  - SDG 5: Gender Equality
- **Planet**: Protecting Earth's ecosystems and promoting responsible resource use to safeguard environmental sustainability:
  - SDG 6: Clean Water and Sanitation
  - SDG 12: Responsible Consumption and Production
  - SDG 13: Climate Action
  - SDG 14: Life Below Water
  - SDG 15: Life on Land
- **Prosperity**: Furthering healthy economic growth, innovation and decent work in ways that respect the planet:
  - SDG 7: Affordable and Clean Energy
  - SDG 8: Decent Work and Economic Growth
  - SDG 9: Industry, Innovation and Infrastructure
  - SDG 10: Reduced Inequalities
  - SDG 11: Sustainable Cities and Communities
- **Peace**: Supporting inclusive, peaceful and just societies through strong institutions and legal frameworks:
  - SDG 16: Peace, Justice and Strong Institutions
- **Partnership**: Mobilizing governments, industries and civil society to cooperate and work together as well as share accountability at a global and local level:
  - SDG 17: Partnerships for the Goals

Each P corresponds with specific SDGs and aligns with the key dimensions of development (UN, 2015). This enables stakeholders to navigate the complexity of the 2030 Agenda without losing sight of its overarching goals.

Figure 2.2 The Sustainable Development Goals with the UN emblem

United Nations

The Sustainable Development Goals: The operational core

The 17 SDGs provide the operational structure of the 2030 Agenda. Figure 2.2 shows the 17 SDGs and the wide range of issues they address. Each goal is represented by a coloured icon containing its number and title of the goal. Each goal has a statement of intent, specific targets and indicators to track progress towards its achievement. There are 169 targets and over 230 indicators in total.

Each goal addresses a critical area of concern: collectively they aim to advance human development and environmental stewardship.

Importantly, the SDGs are interlinked as progress in one goal supports progress in others, and failure in one can undermine the progress of several others.

For example, improving education (SDG 4) enhances health outcomes (SDG 3) and increases economic opportunity (SDG 8), but continuing with unsustainable production and consumption patterns (SDG 12) can jeopardize climate action (SDG 13) and biodiversity (SDG 14 and SDG 15). Understanding these connections is central to making effective policy and operational business decisions.

The UN has identified SDG 4 and SDG 17 as the cross-cutting, facilitative SDGs for enabling progress towards the others. Partnerships facilitate the sharing of knowledge, expertise, resources and best practices, which are essential for accelerating progress towards a sustainable future. SDG 17 demonstrates that achieving sustainable development requires collective action and inclusive partnerships at local, national, regional and international levels. To enable meaningful, constructive partnerships requires education of all. This is clearly demonstrated by SDG 4 Quality Education whose goal is to 'ensure inclusive and equitable quality education and promote *lifelong learning* opportunities for all' (UN, .n.d. b, authors' emphasis).

It is evident that the SDGs serve a critical role in supporting the four pillars of sustainability. As previously outlined, the SDGs provide a global roadmap for action. However, this framework can appear abstract or removed from the practicalities of daily operations. To make the goals more accessible and actionable; it is helpful to view them as a set of tools designed to support the realization of the goals. To explore this further, a closer examination of an individual SDG is useful.

Sustainable Development Goal: SDG 4 Quality Education

SDG 4 was chosen for closer examination because education is the foundation for achieving all other SDGs (UNESCO, 2015). For an SDG to function as a practical tool, knowing how to 'read' their design is useful. Figure 2.3 deconstructs SDG 4 Quality Education into its component parts.

Figure 2.3 SDG 4 Quality Education

EXAMINATION OF SDG 4 QUALITY EDUCATION

- Every SDG has its own icon:
    - The icon for SDG 4 shows that quality education is required for sustainable development.
- Each SDG has a goal:
    - The goal for SDG 4 is 'Ensure inclusive and equitable quality education and promote lifelong learning opportunities for all'.
    - The important word here is the verb 'ensure'. To ensure signifies a commitment to achieving the goal. It requires proactive efforts to 'promote' and accountability to ensure progress towards the targets and indicators.
- Each goal has two sets of targets:
    - SDG 4's targets are: 4.1–4.7 and 4.a–4.c.
    - The 'target' is a specific and measurable objective set to track progress towards achieving the goal.
- SDG 4 has seven outcome targets and three process targets:
    - Outcome targets structure the content of the SDG and are indicated by number, e.g. 4.1, 4.2, 4.3.
    - Process targets increase the 'how to' operationality of the SDG. They are indicated by letters, e.g. 4.a, 4.b, 4.c.
- Each target has indicator(s):
    - SDG 4's indicators for the outcome target are 4.1.1–4.7.1.
    - SDG 4's indicators for the process targets are 4.a.1–4.c.1
    - The indicators aim to facilitate evidencing the progress being made towards a target.

A close analysis of SDG 4 illustrates the implications of the goals for THE. Target 4.7. calls for all individuals to acquire the knowledge, skills, values and attitudes needed to support sustainable development. It emphasizes the importance of lifelong learning extending education beyond formal educational institutions to include professional development. Achievement of work towards outcome Target 4.7 is indicated by '4.7.1 Extent to which (i) global citizenship education and (ii) education for sustainable development are mainstreamed in (a) national education policies; (b) curricula; (c) teacher education and (d) student assessment' (UN, .n.d.).

Importantly, this exact wording is repeated in Indicator 12.8.1 under SDG 12 Responsible Consumption and Production, and Indicator 13.3.1 under SDG 13 Climate Action. This clearly demonstrates that education is a cross-cutting enabler of sustainable behaviour and systems change. The repetition reinforces the message that achieving environmental, social and economic sustainability depends upon widespread education, awareness and capacity building. For THE industries this has clear implications. THE organizations are in a position to promote sustainable behaviour through visitor education initiatives, embed sustainability principles in staff development programmes, and partner with local education providers to strengthen skills aligned with the SDGs.

What is immediately evident from this explanation is that the SDGs are not only impactful visuals for showcasing a THE organization's sustainability credentials, but require close reading to fully appreciate their operational value. However, care must be taken when using them to represent an industry's sustainability credentials.

While the SDGs are intended to guide global efforts towards sustainability, they can sometimes be misused or misrepresented, which can lead to greenwashing. Greenwashing refers to the practice of portraying an organization, product or service as environmentally friendly and conforming to sustainable practices when it does not meet the necessary standards. As part of greenwashing, the SDGs are often used to indicate sustainability credentials to suggest the organization is contributing to the goal, its targets and indicators. Greenwashing undermines the credibility of the SDGs as well as authentic sustainability efforts, eroding trust between businesses, consumers, communities and other stakeholders. To avoid greenwashing, it is important for organizations to demonstrate genuine commitment to sustainability by aligning their actions with the principles of the SDGs, and provide transparent, evidence-based reporting on their sustainability performance.

ACTIVITY 2: APPLYING SDG 12 IN THE

Background

SDG 12 Responsible Consumption and Production aims to 'ensure sustainable consumption and production patterns' (UN, 2015).

This activity focuses on two important targets:

- Target 12.3: Halve per capita global food waste at the retail and consumer by 2030.
- Target 12.8: By 2030, ensure that people everywhere have the relevant information and awareness of sustainable development and lifestyles in harmony with nature.

Instructions

Target 12.3 – Reducing Food Waste

1 **Read Target 12.3 and Indicator 12.3.1** which measures food loss and waste levels.
2 **In the table below**:
   o List actions a food and beverage manager working with an executive chef could plan for and implement to reduce food waste.
   o Consider practical actions they could take with their teams.
   o Suggest methods they could use to measure the success of these actions.

Think about menu planning, food ordering systems, waste audits, kitchen operations.

Target 12.3 Actions	Indicator 12.3.1 Measurement

Target 12.8 – Promoting Awareness

1 **Read Target 12.8 and Indicator 12.8.1** which focuses on education for sustainable development.
2 **In the table below**:
   o List actions a hotel manager or events manager could plan and implement to promote sustainable consumption and awareness among guests, delegates, staff and attendees.
   o Suggest methods to measure the effectiveness of these actions.

Think about guest information, guest feedback surveys, signage, workshops, team training.

Target 12.8 Actions	Indicator 12.8.1 Measurement

REFLECTIVE QUESTION

Reflecting on the link between the two targets, how could THE professionals use education and awareness strategies (SDG 12.8) to bring about mindset and behavioural change that reduces food waste (SDG 12.3)?

2017: Key competencies for sustainability

By 2017, international discussions on sustainability had shifted toward the need for more systemic transformation. In response, UNESCO introduced a framework of key competencies considered essential for achieving the SDGs. These competencies reflected a broader understanding that sustainable development requires not only technical responses but also changes in how individuals think, interact and act in complex and interconnected systems.

For the THE sectors, these competencies offer a structured way to consider the skills and capacities needed to engage with sustainability issues across different operational and cultural contexts. They serve as a foundation for building professional capabilities that support decision-making, and action aligned with long-term sustainability objectives.

The framework is organized around three learning domains: cognitive (ways of thinking), socio-emotional (ways of being), and behavioural (ways of practicing). Each domain addresses a different but interconnected dimension of learning and professional practice:

- The cognitive domain focuses on systems thinking, anticipation and critical reflection.
- The socio-emotional domain includes competencies such as ethical reasoning, self-awareness and intercultural sensitivity.

- The behavioural domain centres on the ability to act strategically, collaboratively and adaptively within real-world challenges.

While the competencies introduced here provide an overview, they are discussed in greater detail in Chapter 3, where their implications for professional roles, organizational practice and sectoral change are explored more fully. This overview provides an initial understanding about how competence-based approaches are increasingly shaping sustainability efforts within THE.

2018–26: Next Tourism Generation Alliance and Pact for Next Tourism Generation Skills (PANTOUR)

The Next Tourism Generation Alliance (NTG) and the Pact for Next Tourism Generation Skills (PANTOUR) are EU-funded initiatives established to address structural and emerging skills needs in the tourism sector. NTG ran from 2018 to 2021 while PANTOUR, launched in 2023, continues this work under the EU Pact for Skills framework until 2026.

The NTG Alliance developed the first pan-European Skills Blueprint for the tourism industry, involving 14 partners across eight countries. It focused on building competencies in three core areas – digital, green and social skills – across five sub-sectors: accommodation, food and beverage, visitor attractions, travel and tour operations and destination management. The Blueprint emphasized the necessity of collaboration between education providers, industry, government and social partners to identify and address skills gaps, especially in light of sector-wide disruptions such as the Covid-19 pandemic.

To support these goals, NTG introduced tools including the Skills Assessment Methodology (SAM), the NTG Skills Matrix and LAB, and a Tourism Sector Skills Toolkit. These instruments were designed to assess current and future workforce needs, align training provision and support strategic skills planning. A Quality Skills Standards Framework was also established to promote coherence between industry needs and education standards at national and EU levels.

PANTOUR builds on this foundation, with a focus on practical implementation of skills strategies at national and regional levels. It aims to improve lifelong learning, support labour mobility and anticipate future skills needs, particularly those linked to green and digital transitions. It leverages collaborative governance mechanisms developed through NTG, including National and Regional Skills Partnerships (N/RSPs), which serve as key vehicles for aligning policy, training provision and industry demand.

Together, NTG and PANTOUR represent a coordinated effort to develop a future-oriented, adaptable and sustainable tourism workforce in Europe.

2021–25: EU Transition Pathway for Tourism and the Sustainable EU Tourism project

The European Union's evolving tourism policy framework reflects a deliberate shift from strategic vision to practical implementation, particularly in the context of environmental, digital, and resilience transitions. Two key initiatives exemplify this trajectory: the Transition Pathway for Tourism and the Sustainable EU Tourism – Shaping the Tourism of Tomorrow project (European Commission, 2022, 2023).

Launched in February 2022, the Transition Pathway for Tourism has provided a structured approach to guiding the sector through major systemic changes. Developed through extensive stakeholder consultation, it identifies 27 action areas critical to achieving a more sustainable and resilient THE sector. These include measures for digital transformation, environmental management, data sharing and improved governance structures. The Transition Pathway is not a regulatory instrument but a reference tool to support coordinated action across public and private actors at EU, national and local levels.

Building directly on the Transition Pathway, the Sustainable EU Tourism project ran from 2023 to 2025. It translated the strategic objectives of the Pathway into applied support for destinations. The project assisted over 200 destinations in identifying specific challenges and implementing context-appropriate responses. A notable feature was the twinning and peer-learning model, which connected destinations with similar priorities to enable practical knowledge exchange and accelerate local-level adaptation.

The developmental relationship between these two initiatives illustrates a broader shift within EU tourism governance – from conceptual planning to facilitated action. This structured progression supports the alignment of strategic ambitions with operational capacity, encouraging destinations to move beyond isolated sustainability efforts toward integrated and collaborative approaches. For professionals entering the THE sector, understanding this policy landscape is essential. It defines the current and future operational environment, shapes funding and regulatory priorities, and sets clear expectations for performance across environmental, social and economic dimensions.

2022: GreenComp, The European Sustainability Competence Framework

GreenComp, the 2022 European Sustainability Competence Framework, provided a structured model for developing competences that support sustainability across multiple sectors. Developed by the European Commission, it outlined 12 competences organized under four interconnected headings:

- **Embodying sustainability values:** Living and making decisions in ways that reflect core sustainability principles like respect for nature, equity and responsibility.
- **Embracing complexity in sustainability:** Recognizing and navigating the interconnected, dynamic systems and trade-offs involved in sustainable development.
- **Envisioning sustainable futures:** Imagining and articulating positive, long-term scenarios that guide transformative change toward sustainability.
- **Acting for sustainability:** Taking informed and purposeful actions – individually and collectively – to promote environmental integrity, social justice and economic wellbeing.

These categories reflected the interwoven cognitive, practical and ethical dimensions of sustainability, positioning the framework within a wider landscape of competence-based approaches to sustainability learning and practice.

GreenComp drew conceptually from the UNESCO Key Competencies for Sustainability, which defined broad, globally-oriented learning outcomes for education for sustainable development. While UNESCO's framework remains normative and international in scope, GreenComp translated these principles into a European context, offering a more operational structure suited to regional policy frameworks such as the European Green Deal and the New Skills Agenda. The shared emphasis on systems thinking, critical reflection, collaboration and anticipatory competence ensures conceptual coherence between the two frameworks.

Table 2.2 illustrates how GreenComp operationalizes UNESCO's key competencies for sustainability.

In THE industries, GreenComp supports the integration of sustainability into strategic and operational domains. These industries, shaped by complex environmental and socio-economic interdependencies, face specific challenges related to resource use, community impact and regulatory compliance. GreenComp provides a reference point for aligning sectoral practices with wider sustainability transitions, contributing to the development of context-sensitive approaches that are both locally grounded and globally informed.

Table 2.1 The alignment of GreenComp competences with UNESCO's key competencies for sustainability

UNESCO competency	UNESCO competency objective	Corresponding GreenComp competence area	Related GreenComp competences
Systems thinking	Understand and reflect upon interconnections within and between systems	Embracing complexity on sustainability	Systems thinking; Critical thinking
Anticipatory thinking	Understand and evaluate multiple futures; plan and act with foresight	Envisioning sustainable futures	Futures literacy; Adaptability
Critical thinking	Question norms, practices, and opinions; reflect on one's own values and assumptions	Embracing complexity on sustainability	Critical thinking; Systems thinking
Normative	Understand and reflect on norms and values; assess sustainability values and goals	Embodying sustainability values	Valuing sustainability; Supporting fairness
Self-awareness	Reflect on own role, values and ability to influence change	Embodying sustainability values	Self-awareness; Valuing sustainability
Strategic	Develop and implement innovative actions to progress sustainability goals	Acting for sustainability	Political agency; Collective action
Collaboration	Learn from others; understand and respect diverse perspectives; work in partnerships	Acting for sustainability	Collective action; Political agency
Integrated problem-solving	Apply multiple competencies to solve complex sustainability patterns	Cross-cutting across all GreenComp areas	All 12 competencies contribute to integrated problem-solving

Adapted from UNESCO (2017); GreenComp (2022)

Table 2.2 Timeline of key milestones in global sustainability and their relevance to THE

Year	Milestone
1972	**Stockholm Conference and United Nations Environment Programme (UNEP) held.** These events marked the start of coordinated international efforts to manage environmental impacts – laying a foundation for tourism's environmental accountability and the basis for responsible tourism practice.
1975	**United Nations World Tourism Organization (UNWTO) formed.** UNWTO's creation formalized global tourism policy leadership – key for integrating sustainability in tourism development. Now operating as UN Tourism.
1987	**The Brundtland Report: *Our Common Future* published.** Introduced most widely used definition of sustainable development and laid the conceptual groundwork for the three pillars of sustainability – economic, environmental and social – which underpin THE sustainability models.
1989	**The Hague Declaration on Tourism released.** The first key policy linking tourism to sustainability; it emphasized tourism's dependence on environmental and cultural resources.
1992	**The Earth Summit's Agenda 21 published.** It introduced sustainable tourism as a global priority – a guiding framework still used in destination planning. It is considered an action plan for sustainable tourism and local engagement.
1994	**Elkington's Triple Bottom Line theory presented.** The concept of People, Planet, Profit has become central to assessing tourism's sustainability performance.
2001	**UNESCO's Universal Declaration on Cultural Diversity announced.** Culture as a pillar of sustainability was introduced.
2003	**UNESCO's Convention for the Safeguarding of the Intangible Cultural Heritage published.** The importance of cultural tourism as a sustainability issue was reinforced.
2005	**UNESCO's Convention on the Protection and Promotion of the Diversity of Cultural Expressions ratified.** The sustainability of the principles of cultural diversity was legally reinforced.
2010	**Pillars of Sustainability Framework revised.** The recognition of culture as the fourth pillar of sustainability is introduced. The model provides a lens for sustainability planning in the sector. The framework remains especially relevant for THE industries.
2015	**UN 2030 Agenda for Sustainable Development and the Sustainable Development Goals (SDGs) released.** The SDGs provided a shared language for all sectors – including THE – to align with global sustainability targets. It is considered a key global roadmap for sustainability in THE.

(continued)

Table 2.2 (Continued)

Year	Milestone
2017	**UNESCO's Key competencies for sustainability introduced.**
	UNESCO introduced eight core competencies for sustainability organized around ways of thinking, ways of doing and ways of being. The key competencies established a global standard for sustainability education – now essential for THE education and professional development.
2018	**European Commission's The Next Tourism Generation Alliance (NGTA) research initiative launched.**
	The NTGA research initiative focused on identifying future skills needed for THE industries to develop frameworks for improving education and training provision for THE.
2021	**European Commission's Transition Pathway for Tourism: Leading the Transition launched.**
	The transition pathway outlined a strategic framework co-created with THE stakeholders to guide the sector's green and digital transformation.
2022	**GreenComp: The European Sustainability Competence Framework.**
	This framework was designed to connect education with the professional sustainability skills needed in THE.
2023	**Pact for Next Tourism Generation Skills (PANTOUR).**
	Continued the work of the NGTA, focusing on closing the tourism skills gap through digital, green and social innovation. PANTOUR expands this effort across Europe by aligning the skills with the THE sector's sustainability goals.
2023	**European Union's Sustainable EU Tourism: Shaping the Tourism of Tomorrow project commenced.**
	The project aimed to support tourism destinations in adopting sustainable practices through collaboration, knowledge exchange and alignment with the EU's Transition Pathway for Tourism.
2024	**The Statistical Framework for Measuring the Sustainability of Tourism (SF-MST).**
	Enabled consistent sustainability reporting across tourism destinations for transparency and informed decision-making. It provided a common language for evaluating and reporting tourism impacts.

2024: The Statistical Framework for Measuring the Sustainability of Tourism (SF-MST)

The Statistical Framework for Measuring the Sustainability of Tourism (SF-MST), developed by the UN Statistics Division and UNWTO, provided an internationally agreed methodology for integrating sustainability into tourism statistics. The SF-MST framework continues to enable countries to assess tourism's economic, environmental and social dimensions in a coherent and comparable manner.

The SF-MST framework recognizes that conventional tourism statistics, which focus primarily on visitor numbers and economic contributions, are insufficient for understanding the full implications of tourism activity. By linking environmental indicators (such as resource use, emissions and land use) and social aspects (including employment, cultural heritage and community wellbeing) to tourism data, the framework supports a more complete understanding of tourism's impacts and dependencies.

For THE industries, the SF-MST provides a statistical basis for aligning with broader sustainability agendas, including the SDGs and national climate and biodiversity targets. It also facilitates evidence-based policy and planning by offering a consistent structure for tracking trade-offs and synergies between tourism development and sustainability objectives.

The SF-MST framework represents a key advancement in the standardization of sustainability-related tourism data. It supports national statistical systems and industry stakeholders in measuring and managing tourism in ways that reflect environmental limits and social priorities. It is designed for gradual implementation, recognizing differing national capacities and data systems, while providing a reference point for future integration of sustainability into tourism governance and decision-making.

Conclusion

As Table 2.1 demonstrates, the history of sustainability in the THE sectors reveals a clear path: from early global declarations and high-level strategies to the development of frameworks that now guide real-world practice. Key events such as the 1972 Stockholm Conference, the 1987 Brundtland Report and the adoption of the 2030 Agenda for Sustainable Development established the ethical and strategic groundwork for linking THE with sustainable development.

More recent developments mark a shift from policy and planning to implementation and accountability. Initiatives such as the Statistical Framework for Measuring the Sustainability of Tourism (SF-MST), GreenComp and the Pact for Next Tourism

Generation Skills (PANTOUR) illustrate this transition. These tools support the integration of sustainability into operational practice by defining professional competencies, measuring sectoral impacts, and linking THE activity directly to the Sustainable Development Goals (SDGs).

For emerging professionals in THE, this means preparing not only through theoretical knowledge but by developing applied capabilities aligned with established frameworks. Central to this preparation are the UNESCO Key Competencies for Sustainability which are now recognized as essential for professionals operating in complex, interconnected environments. These competencies underpin the development of green skills, digital skills and social skills identified by NTG and PANTOUR.

Sustainability in THE is no longer a policy aspiration; it is a practical requirement. Future professionals must be ready to interpret sustainability frameworks, engage with measurable standards and lead change across diverse contexts. Understanding the historical development of these frameworks is key to making sense of their current relevance and to applying them responsibly in professional practice.

KEY TAKEAWAYS

- Historical foundations of sustainability in THE:
    - Key milestones such as the 1972 Stockholm Conference, the 1987 Brundtland Report and the 2015 UN 2030 Agenda laid the groundwork for integrating sustainability into THE policy and practice globally.
- THE's role in global sustainability:
    - THE sectors and its stakeholders are positioned as critical contributors to achieving the SDGs.
    - THE sectors need to develop inclusive partnerships, shared governance and lifelong education to achieve sustainability.
- Four pillars of sustainability and the Triple Bottom Line (TBL):
    - Sustainability in THE is structured around four interconnected pillars: environmental, social, economic and cultural. Weakness in one affects the overall sustainability framework, highlighting the need for a balanced approach.
    - The TBL is a guiding model in THE for assessing sustainability performance across different stakeholder groups, emphasizing the need for holistic stakeholder engagement.
- Cultural sustainability is a core dimension:
    - UNESCO conventions and declarations have affirmed culture as a fundamental element of sustainable development, encouraging THE professionals to support living cultural heritage and avoid cultural commodification.

    - There is a legal obligation to safeguard cultural heritage.
- Competency-based approaches for sector transformation:
    - Frameworks such as UNESCO's key competencies, GreenComp and the PANTOUR initiative emphasize the need for future-oriented, systems-based thinking across THE professions.
- From policy to practice – applied sustainability tools:
    - Instruments like the Statistical Framework for Measuring the Sustainability of Tourism (SF-MST) enable countries and organizations to assess and monitor sustainability impacts through standardized, actionable metrics.

REFLECTIVE QUESTIONS

1 What does sustainability mean in THE, and why is it important for all THE industries?
2 What are some of the ways you can protect local culture and support communities while still helping a business to succeed?
3 Which frameworks and instruments will become increasingly influential on THE operations in the future? Why?
4 Which skills and competencies do you think you will need to help solve the sustainability problems in THE? How can you start building them now?

Further resources

Explore the UN Tourism website to get a better understanding of the work this global organization is doing to ensure a sustainable future for THE industries: www.unwto.org (archived at https://perma.cc/SAH4-QH8E)

Become familiar with the skills development frameworks:

- UNESCO's key competencies: https://unesdoc.unesco.org/ark:/48223/pf0000245056 (archived at https://perma.cc/LT8M-VKBV)
- PANTOUR: https://nexttourismgeneration.eu/tourism-sector-and-skills-toolkit/ (archived at https://perma.cc/G5AN-VAXJ)
- GreenComp: https://joint-research-centre.ec.europa.eu/greencomp-european-sustainability-competence-framework_en (archived at https://perma.cc/K5ER-ESRB)

References

Amos, J (2017) Kaikoura: 'Most complex quake ever studied', BBC News, 23 March, www.bbc.co.uk/news/science-environment-39373846 (archived at https://perma.cc/3W7Q-6YBH)

Eco-Business (2014) EarthCheck awards NZ's Kaikoura Community for 10 years of sustainable tourism, Eco-Business, 7 April, www.eco-business.com/press-releases/earthcheck-awards-nzs-kaikoura-community-10-years-sustainable-tourism (archived at https://perma.cc/U9EM-JGGA)

Elkington, J (2004) Enter the Triple Bottom Line, 17 August, https://johnelkington.com/archive/TBL-elkington-chapter.pdf (archived at https://perma.cc/4WLP-25ZT)

European Commission (2022) Transition Pathway for Tourism, Publication Office of the European Union, https://data.europa.eu/doi/10.2873/344425 (archived at https://perma.cc/Y9R8-PVNP)

European Commission (2023) Sustainable EU Tourism – Shaping the Tourism of Tomorrow, https://transport.ec.europa.eu/tourism/transition-eu-tourism/sustainable-eu-tourism-shaping-tourism-tomorrow_en (archived at https://perma.cc/5T7F-7GNC)

Hall, L (2025) The seven travel trends that will shape 2025, BBC Travel, 6 January, www.bbc.co.uk/travel/article/20250106-the-seven-travel-trends-that-will-shape-2025 (archived at https://perma.cc/E6XG-LHHQ)

Institute of Civil Engineers (.n.d.) Moving mountains to reconnect communities, www.ice.org.uk/what-is-civil-engineering/infrastructure-projects/moving-mountains-to-reconnect-communities (archived at https://perma.cc/F979-VX7C)

Moorfield, J C (2011) *Te Aka Māori-English, English – Māori Dictionary* (3rd ed), Pearson, New Zealand

Neeraj, S, Mannakkara, S and Wilkinson, S (2021) Evaluating socio-economic recovery as part of building back better in Kaikoura, New Zealand, *International Journal of Disaster Risk Reduction*, 52, pp 1–10, https://doi.org/10.1016/j.ijdrr.2020.101930 (archived at https://perma.cc/V724-VNYR)

Night Sky Tourist (2019) What is AstroTourism? https://nightskytourist.com/astrotourism (archived at https://perma.cc/DC7Y-P7RQ)

PANTOUR (2022) Plan for sectoral cooperation to address skills needs in the tourism sector, https://nexttourismgeneration.eu/blueprint-for-sectoral-cooperation-to-address-skills-needs-in-the-tourism-sector (archived at https://perma.cc/5XZ3-YBWB)

PANTOUR (2023) Pact for Next Tourism Generation Skills, https://nexttourismgeneration.eu/pantour/ (archived at https://perma.cc/Y9XZ-V5HG)

RTNZ (2025) The community driving Kaikoura's destination management journey, YouTube, 11 March, www.youtube.com/watch?v=8cmrfm6vVDc (archived at https://perma.cc/AX5Y-ZSMB)

Statistics New Zealand (2018) 2018 Census population and dwelling counts, www.stats.govt.nz/information-releases/2018-census-population-and-dwelling-counts/#update (archived at https://perma.cc/45RE-KSCR)

Tarantino, J (2023) Embarking on ethical adventures: Whale watching in Kaikoura, New Zealand, The Environmental Blog, 17 August, www.theenvironmentalblog.org/2023/08/embarking-on-ethical-adventures-whale-watching-in-kaikoura-new-zealand (archived at https://perma.cc/DY36-W9J8)

Thusoo, S, Cordova-Arias, C, Jeong, S, Aigwi, E, Vega, E and Feinstein, T (2019) Improving earthquake resilience through community engagement, Learning from Earthquakes, https://lfestorage.s3.us-east-2.amazonaws.com/images/Activities/Travel_Study/2019_Travel_Study/Final_Report_Socio_Economic_Team.pdf (archived at https://perma.cc/S7DH-XN6S)

UN (1992) United Nations Conference on Environment & Development: Agenda 21, 3-14 June, https://sustainabledevelopment.un.org/content/documents/Agenda21.pdf (archived at https://perma.cc/TEE3-KEPG)

UN (2015) Transforming Our World: The 2030 Agenda for Sustainable Development. New York: United Nations, https://docs.un.org/en/A/RES/70/1 (archived at https://perma.cc/7JJ4-6WWW)

UN (.n.d. a) United Nations Conference on the Human Environment, 5-16 June 1972, www.un.org/en/conferences/environment/stockholm1972 (archived at https://perma.cc/M35A-4TPY)

UN (.n.d. b) Sustainable Development Goals, www.un.org/sustainabledevelopment (archived at https://perma.cc/YG8G-BCDT)

UNESCO (2001) Universal Declaration on Cultural Diversity, www.unesco.org/en/legal-affairs/unesco-universal-declaration-cultural-diversity

UNESCO (2003) Convention for the Safeguarding of the Intangible Cultural Heritage, www.unesco.org/en/legal-affairs/convention-safeguarding-intangible-cultural-heritage

UNESCO (2005) Convention on the Protection and Promotion of the Diversity of Cultural Expressions, www.unesco.org/creativity/en/2005-convention

UNESCO (2017) Education for Sustainable Development Goals Learning Objectives, https://unesdoc.unesco.org/ark:/48223/pf0000247444 (archived at https://perma.cc/T6W9-UDMB)

UNESCO (2023) Concepts Glossary: Culture, www.unesco.org/interculturaldialogue/en/concept-glossary#:~:text=Culture

UN Tourism (2024) Statistical Framework for Measuring the Sustainability of Tourism (SF-MST), www.unwto.org/tourism-statistics/statistical-framework-for-measuring-the-sustainability-of-tourism (archived at https://perma.cc/3HMM-8TFM)

UN Tourism (.n.d.) About UN Tourism, www.unwto.org/who-we-are (archived at https://perma.cc/D3MG-H8UF)

UNWTO (1989) The Hague Declaration on Tourism, 2 June, https://digitallibrary.un.org/record/63652?v=pdf (archived at https://perma.cc/9WTN-U86H)

UNWTO (2008) Tourism Satellite Account: Recommended Methodological Framework 2008, www.unwto.org/tourism-statistics/on-economic-contribution-of-tourism-tsa-2008 (archived at https://perma.cc/B37Z-BP5V)

Whale Watch Kaikoura (.n.d.) Kaitiakitanga / Conservation, https://whalewatch.co.nz/our-home/kaitiakitanga-conservation (archived at https://perma.cc/77VR-E5NQ)

World Commission on Environment and Development (1987) *Our Common Future*, Oxford University Press, Oxford

Žerajić, A (2020) Multilingualism in tourism: Tourism in function of development of the Republic of Serbia, *Tourism International Scientific Conference Vrnjačka Banja – TISC*, 5 (1), pp 518–35, www.tisc.rs/proceedings/index.php/hitmc/article/view/356 (archived at https://perma.cc/UAW6-XULV)

3 | Professional development for sustainability

  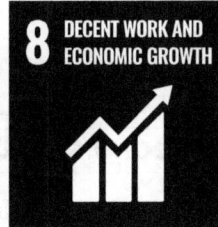 

| Critical thinking competency | Strategic competency | Self-awareness competency | Normative competency |

CHAPTER AIM

The aim of this chapter is to explore the role of professional development in building sustainability competency within tourism, hospitality and events (THE) industries. Framed by SDG 4: Quality Education (UN, .n.d.), the chapter introduces how professional growth can be shaped through structured learning models. The chapter also connects to SDG 5 and SDG 8 through considerations of equal opportunities, and inclusive and fair workplace practice. Through engagement with UNESCO's key competencies for sustainability and the OECD Learning Compass, learners are supported to examine the relationship between personal agency, organizational training needs and the evolving expectations of professionalism in THE.

The chapter supports learners to develop the four key competencies:

- Critical thinking competency – to evaluate the quality and relevance of sustainability knowledge and reflect upon the implications of professional learning choices.
- Self-awareness competency – to recognize personal motivations, values and goals, and how these shape professional development pathways.
- Normative competency – to understand and navigate the ethical and value-laden dimensions of sustainability in decision-making and leadership.
- Strategic competency – to plan and implement sustainability-oriented development initiatives within individual roles and organizational structures.

LEARNING OUTCOMES

Upon completion of this chapter, you will be able to:

- explain the relevance of sustainability to professional development in the context of THE industries
- critically evaluate your current competencies and identify priority areas for growth using sustainability frameworks
- apply self-awareness and normative reasoning to reflect on values and ethical responsibilities in your professional practice
- develop a context-specific sustainability training or development plan using strategic tools such as training needs analysis and the Learning Compass
- understand how structured professional development supports long-term sustainability outcomes for both individuals and organizations

KEY WORDS

Sustainability competencies, professional development, Learning Compass, training needs analysis, self-awareness, values-based leadership, strategic planning, critical thinking.

Introduction

THE industries are increasingly shaped by complex global dynamics that include climate instability, biodiversity loss, social inequality and shifting labour markets. These conditions directly impact the sectors operations and future viability. In response, many governments, businesses and international organizations are calling for a transition to more sustainable economic models. For THE, this transition requires more than adopting new technologies or adjusting business models; it calls for a transformation in how the sector understands its responsibilities and how its professionals are prepared to meet them.

The growing emphasis on sustainability reflects a broader recognition that industries like THE are embedded within ecological and social systems, and that their success depends on how they manage their relationship with these systems. Pressures to reduce carbon emissions, manage resources more efficiently and demonstrate social accountability are now part of the regulatory and operational environment in many destinations. Alongside these pressures is a growing public expectation that THE organizations act ethically and equitably, particularly when working with communities, cultures and environments that are not their own.

At the same time, the internal dynamics of the THE workforce present their own challenges. High staff turnover (Gabbitas, 2023), short-term employment models and skills mismatches (Manpower Group, 2024) continue to characterize many areas of the industry. These issues undermine the continuity and depth of professional practice needed to support sustainability transitions. They also reflect a wider pattern in which training and development are often reactive, inconsistent or disconnected from the values that sustainability requires.

In this context, the question of how professionals in THE are developed – what they learn, how they are supported, and what kind of capacities they cultivate – becomes central. It raises critical issues about education and training not just as technical instruction, but as a means of preparing individuals to engage thoughtfully and effectively with the complexity of the world.

The idea of professionalism in THE must therefore evolve. It must account for the need to collaborate with diverse stakeholders, to navigate ethical dilemmas, to make decisions under uncertainty, and to think long-term about the social and environmental consequences of everyday business practices. These are not abstract concerns; they are becoming the basis on which organizations are judged by investors, regulators, communities and customers alike.

In this shifting environment, sustainability is not an added responsibility. It is becoming central to how THE operates and how its people are expected to think, relate and act. The ability to develop and apply the relevant competencies is now essential to both individual career progression and organizational resilience. Thus, there is a growing need to embed sustainability into the core of professional development across the sector. As the Manpower Group (2024) state, 'By 2030, the green transition is expected to create up to 30 million new jobs… Business leaders must close the [talent] gap by helping workers develop the skills they need for the green jobs of the future.' To effect the transformation for sustainability, a mindset change in the THE industry is required.

Developing a sustainability mindset

Meeting sustainability challenges in THE requires more than technical competence; it involves a shift in how individuals perceive their role in relation to environmental and social systems. This shift is often described as the development of a *sustainability mindset*, a way of thinking, being and behaving that reflects an ongoing commitment to sustainability values.

Rimanoczy (2020) defines the sustainability mindset as an awareness of the interdependence between humans and the environment, coupled with a sense of responsibility for contributing to a more sustainable world. It involves curiosity, systemic thinking and a commitment to ethical action. It is not a set of instructions or fixed behaviours, but a lens through which professionals interpret their choices and challenges. In the context of THE, this means recognizing that everyday business practices such as sourcing materials, engaging with host communities or managing energy usage are not neutral acts, but decisions with ethical and ecological consequences.

This understanding aligns with the concept of the growth mindset (Dweck, 2015) that emphasizes the capacity for learning and change. A growth mindset supports the idea that sustainability is not a state to be achieved, but a continual process of reflection and development. Dweck's idea of 'not yet' is particularly relevant here. Rather than viewing gaps in knowledge or failed attempts as signs of inability, they are seen as part of an evolving process. This framing encourages professionals to persist in the face of complexity and to treat challenges as opportunities to adapt and grow.

The need for such a mindset is increasingly recognized at the policy level. In his opening remarks to the 36th joint meeting of the UN Tourism's Commission for Asia and the Pacific (CAP) and its Commission of South Asia (CSA), the Philippines President Ferdinand Marcos Jnr emphasized that transformation in tourism requires education that elevates not only skills but values. He warned that tourism, if unmanaged, can harm local cultures and environments, and called for training that prepares professionals to engage meaningfully and respectfully with the communities they serve:

> 'Education is always going to make any industry better and certainly tourism is no different. We need to raise the standards and practices in this crucial sector by investing in *education,*

training, and skills upgrading of all the personnel who are working in this industry. We will require experts and professionals from different fields to come and help to make tourism a meaningful and educational and impactful experience, not only for our tourists, but also for our stakeholders. For what good will any tourism effort be if it ends up destroying the local culture and ecology?' (Marcos, in RTVMalacanang, 2024, our emphasis.)

In the THE sector, a sustainability mindset combines the ability to navigate uncertainty with a deeply held awareness of the industry's broader impacts. Benton (Sustainability Stories, 2023) argued that 'most of sustainability is about changing people's minds'. For Rimanoczy (2020) and President Ferdinand Marcos Jnr (RTVMalacanang, 2024) a sustainability mindset is not only desirable, but essential for future-ready professionalism in THE.

SDG 4: Quality Education, Lifelong Learning and Continuous Professional Development (CPD)

Education is a critical foundation for sustainable development (UN, .n.d.). In THE it supports the technical competencies required to deliver services but also the values, behaviours and critical thinking skills needed to engage responsibly with global, national and local challenges. Within this broader understanding of education, SDG 4 Quality Education is highly relevant. It aims to 'ensure inclusive and equitable quality education and promote lifelong learning opportunities for all' (UN, .n.d.).

The term 'lifelong learning' refers to the continuous acquisition of knowledge for personal or professional purposes throughout an individual's life. It includes formal education, non-formal education (such as workplace training or short courses) and informal learning (peer collaboration or experiential learning). THE is an industry characterized by rapid change and evolving global expectations, thus lifelong learning is essential for remaining informed and adaptable.

A key component of lifelong learning in the professional context is continuous professional development (CPD). CPD is the process by which professionals develop and maintain their knowledge and skills related to their professional lives. It is a continuous, proactive and often self-directed form of learning that supports individual development and professional competence (CIPD, .n.d.).

Together, lifelong learning and CPD are commitments to ongoing education and training beyond initial qualifications by engaging in formal and informal learning opportunities. They require participation in conferences, workshops, seminars and professional courses to keep informed about the latest developments, trends and innovations in THE. This extends to networking to build and maintain professional relationships with colleagues, industry experts and organizations.

The importance of lifelong learning and CPD is underpinned by SDG 4 Target 4.7, which requires 'all learners [to] acquire the knowledge and skills needed to promote sustainable development, including, among others, through education for sustainable development and sustainable lifestyles, human rights, gender equality, promotion of a culture of peace and non-violence, global citizenship and appreciation of cultural diversity and of culture's contribution to sustainable development' (UN, .n.d.). The target is echoed in SDG 12 Responsible Consumption and Production and SDG 13 Climate Action.

Importantly, the wording of Indicator 4.7.1 'Extent to which (i) global citizenship education and (ii) education for sustainable development are mainstreamed in (a) national education policies; (b) curricula; (c) teacher education and (d) student assessment' (UN, n.d.) is repeated in SDG 12 and SDG 13 as illustrated in Figure 3.1.

This repetition across the goals reflects the view that sustainability education is not sector-specific but a global expectation across industries. For THE this means that CPD aligned with the goals is not optional or supplementary, but a professional standard that links individual development with broader social and environmental responsibilities.

Figure 3.1 Cross-sector relevance of SDG 4.7.1

SDG 4 Quality Education

Target 4.7: By 2030, ensure all learners acquire the knowledge and skills needed to promote sustainable development…

Repeated in

SDG 12 Responsible Consumption and Production

Target 12.8: By 2030, ensure that people everywhere have the relevant information and awareness for sustainable development and lifestyles in harmony with nature

SDG 13 Climate Action

Target 13.3 Improve education, awareness-raising and human and institutional capacity on climate mitigation, adaptation, impact reduction and early warning

SDG 12 Responsible Consumption and Production Target 12.8 and SDG 13 Climate Action Target 13.3 repeat the required outcome for SDG 4 Quality Education Target 4.7.1.
UN, .n.d.

Grounding sustainability in professional development

As THE industries respond to environmental and social challenges, there is a growing need for approaches to professional development that are not only aligned with global frameworks such as the SDGs but also grounded in values and practices that connect organizations meaningfully to the environment and communities in which they operate. In this context, Fernández (2021) proposes the concept of 'grounding sustainability' as a business principle requiring THE to consider its relationship to how sustainability is understood, embedded and enacted.

Grounded sustainability involves two interlinked processes: returning to the fundamentals and being actionable. For THE, returning to the fundamentals requires a clear and shared understanding of what sustainability means – not only as a concept but as a guiding principle for decision-making. This includes acknowledging THE's dependency on natural ecosystems, cultural integrity and social well-boing, and understanding how its operations directly affect these systems. Rather than viewing sustainability as a checklist of environmental measures, grounding calls for THE organizations to embody sustainability in their strategic planning, governance and daily actions.

Being actionable, the second aspect of Fernández's framework, calls for sustainability commitments to be translated into clear, practical and ethical policies that guide behaviour at all levels of the organization. This means ensuring that professional development initiatives are not generic but intentionally designed to address real-world sustainability issues. It also requires training to align with the four pillars of sustainability and key policy areas to devise meaningful strategies rather than narrow, efficiency-driven outcomes. Professional development, when grounded in sustainability, becomes an approach to improve employee performance by cultivating reflective professionals capable of making ethical and informed decisions. It enables individuals to critically assess the sustainability implications of their roles and to take responsibility for their ongoing learning and growth.

> **INDUSTRY VOICE: MIRANDA SIMMONS**
>
> Miranda Simmons is a freelance sustainable production advisor, consultant, and trainer for media organizations. One of her training clients is BAFTA Albert, the sustainability organization for the film and TV industry. She says: 'BAFTA Albert is the leading organization in the screen industry for environmental sustainability. Albert is promoting professional development for staff, helping build a knowledgeable workforce capable of implementing sustainability measures effectively. A key area of focus is encouraging productions to hire local crews, which can reduce the environmental impact of travel and support regional economies.'

For THE professionals, this approach to development is particularly important. They often operate in diverse cultural and ecological contexts where THE operations can either support or undermine local wellbeing. By grounding sustainability, professionals are better equipped to recognize their ethical responsibilities and navigate competing stakeholder interests.

This grounded approach also ensures that CPD is not a reactive response to external pressures but a proactive commitment to long-term transformation. It provides the foundation for identifying the specific sustainability competencies that professionals need to acquire and apply to their professional role.

To begin this process, THE organizations and professionals might reflect on the following questions as part of their sustainability development strategy:

1 What is the industry's understanding of sustainability? Is it representative of an internationally agreed, global perspective or does it remain operational and fragmented?

2 What skills and competencies do industry need to address sustainability issues? Are current employees equipped to deal with the environmental and social responsibilities embedded in their role?

3 Where can industry access robust frameworks to assess the organization and its employees' training needs? Are sustainability frameworks such as UNESCO's key competencies and the key competencies of sustainability being used to guide staff development?

4 Are current professional development programmes supporting internal and external stakeholders to build sustainability capabilities? How well are current training programmes aligned with sustainability values and practices?

5 How might the industry, organizations and other stakeholders expediate sustainability skills development? What forms of collaboration, resource allocation and strategic planning are needed to accelerate capacity building?

These questions form a practical entry point for grounding sustainability through professional development. They provide the foundation for identifying and applying the key competencies required for sustainable THE practice.

Key competencies for sustainability

Developing a sustainability mindset and grounding professional development in ethical, actionable principles provides a foundation for identifying the specific competencies that individuals and organizations in the THE industries need. These competencies are not arbitrary. They are articulated in internationally recognized frameworks designed to support the implementation of the SDGs particularly SDG 4.7, which emphasizes the importance of education for sustainable development, global citizenship and cultural understanding.

In 2017, UNESCO formalized a set of Key Competencies for Sustainability, reflecting a growing consensus that sustainable development requires more than technical knowledge. It demands the capacity to think critically, relate ethically and act strategically in complex and changing environments. These competencies are organized across three interrelated domains of learning: cognitive, socio-emotional and behavioural.

Cognitive competencies (ways of thinking)

The cognitive domain focuses on the skills necessary to engage with sustainability challenges in a knowledgeable and forward-thinking way. These competencies support professionals in recognizing patterns, anticipating consequences and framing problems within broader systems:

- **Systems thinking competency**: Understanding the interconnectedness of social, environmental and economic systems and recognizing their cause-and-effect relationship.
- **Anticipatory thinking competency**: Envisioning possible futures and preparing for uncertainty and change.
- **Critical thinking competency**: Questioning assumptions and dominant ideas, evaluating evidence and considering multiple perspectives.

In the THE context, these cognitive skills are necessary not only for understanding sustainability issues, but also for framing and reframing them in ways that support innovation and long-term strategy. For example, anticipating the long-term impact of visitor flows on fragile environments requires both systems awareness and the ability to assess policy options critically.

Socio-emotional competencies (ways of being)

This domain emphasizes personal reflection, ethical reasoning and intercultural sensitivity. Ways of being are concerned with values, attitudes and interpersonal capacities and as such are essential in people-centred industries such as THE:

- **Normative competency:** Understanding and navigating ethical dimensions such as fairness and the common good
- **Self-awareness competency:** Recognizing one's positionality, values and biases and understanding how they affect professional decision-making.

For THE professionals, these competencies are central to working in culturally diverse contexts and in roles where decisions directly impact local communities. Socio-emotional learning supports the development of ethical leadership and inclusive, culturally sensitive practice.

Behavioural competencies (ways of practicing)

Behavioural competencies focus on the ability to put knowledge and values into action in professional settings. These are the competencies most closely aligned with organizational outcomes and strategic leadership:

- **Strategic competency:** Designing and implementing action plans that align with sustainability goals.
- **Collaboration competency:** Working constructively with others to solve problems and build partnerships.
- **Integrated problem-solving competency:** Drawing on and combining diverse types of knowledge and skills to respond to multilayered challenges.

These are particularly relevant in THE project design, destination management and operations where collaboration and cross-sector coordination are often necessary.

Together the key competencies for sustainability provide a framework for sustainable professionalism in THE. They reflect what THE professionals must know, how they must relate to others, and how they must behave when addressing complex sustainability challenges. Importantly, they also align with the concept of grounded sustainability (Fernández, 2021) by emphasizing ethics, systems thinking and agency.

ACTIVITY 1: MAPPING YOUR SUSTAINABILITY COMPETENCIES

Purpose

To reflect upon your current capabilities and identify professional development needs using UNESCO's (2017) key competencies for sustainability.

Instructions

1 **Self-assessment**

   Review the sustainability competencies listed below. For each, rate your current level of confidence using this scale:

   o not yet developed

   o developing

   o competent

   o confident

   o expert

Then briefly note one example of how this competency shows up in your experience (e.g. from studies, work, community involvement) and one step you could take to develop it further.

   o Which competencies are most relevant for your future role in THE?

   o How could THE organizations support employees in building these competencies?

Competency	Rating (1 – 5)	Example from experience	Next step for development
Systems thinking			
Anticipatory thinking			
Critical thinking			
Normative competence			
Self-awareness competence			
Strategic competence			
Collaboration competence			
Integrated problem-solving competence			

> **2 Peer discussion (optional)**
>
> In pairs or small groups, share one insight from your self-assessment and discuss how these competencies might differ in priority depending on different roles in the THE sector (e.g. front-line staff vs. sustainability manager).

Extending competency through the OECD Learning Compass

The OECD Learning Compass 2030, part of the Future of Education and Skills 2030 project, offers a forward-looking framework for learning and professional development in response to complex global challenges. While originally designed for educational systems, its relevance extends to professional contexts – particularly for those working in dynamic, people-centred sectors such as THE. The Learning Compass builds upon and extends UNESCO's key competencies for sustainability by introducing a broader view of lifelong learning that encompasses knowledge, skills, attitudes, values and agency (OECD, 2019a).

The framework recognizes that professionals must be prepared not only to respond to change but to shape it. This requires competencies that support individual wellbeing, ethical leadership and systemic transformation. As seen in Figure 3.2, at the heart of the Learning Compass is the idea of learner (or professional) agency – the capacity to set goals, make informed choices and act with purpose. This capacity is reinforced through co-agency, which recognizes the role of peers, mentors, institutions and professional communities in supporting learning (OECD, 2019a). As such, it contributes to the development of a sustainability mindset. As discussed previously, cultivating this mindset is not only about the acquisition of information but also involves reshaping how individuals see themselves, their work and their responsibilities.

Core components of the Learning Compass for professional development

The Learning Compass is comprised of core components. There are four interconnected transformative competencies – knowledge, skills, attitudes and values (OECD, 2019b). These function as the 'points of orientation' for learners to fulfil their potential and contribute to the wellbeing of their communities and the planet.

Moving from the centre outwards, the core foundations represent the essential conditions and baseline capabilities that support all other areas of development. The transformative competencies are those learners need to shape their future as responsible professionals. The competencies are developed through a cyclical learning process to ensure continuous professional development and lifelong learning (EduSkills OECD, 2019).

Core foundations

Core foundations refer to the fundamental conditions and enabling capacities upon which all other aspects of professional development are built. These include cognitive foundations (such as literacy, numeracy and digital literacy), health foundations (encompassing physical and mental wellbeing), and social and emotional foundations (including self-awareness, empathy and moral sensitivity).

For THE professionals, these foundations are critical. Cognitive skills enable informed decision-making; health foundations support resilience in high-pressure environments; and social-emotional foundations encourage inclusivity and ethical engagement.

Knowledge, skills, attitudes and values

Knowledge

Knowledge, as defined in the Learning Compass, includes both theoretical understanding and practical insight. There is an emphasis on being able to understand, interpret and apply knowledge in various scenarios.

The framework identifies four types:

- **Disciplinary knowledge** is considered essential for understanding, and a structure through which learners can develop other type of knowledge.
- **Interdisciplinary knowledge** facilitates combining knowledge from other knowledge bases, and the creation of new knowledge.
- **Epistemic knowledge** involves knowing how to think and behave as a practitioner by understanding how knowledge is constructed. It shows the relevance and purpose of the knowledge.
- **Procedural knowledge** is the understanding of how tasks are performed as a structured process to enable complex problem solving.

This multidimensional approach ensures that professionals in THE can engage with complexity and apply knowledge across varied and evolving contexts (OECD, 2019c).

Skills

Skills refer to the capacity to apply knowledge effectively and responsibly. The Learning Compass distinguishes three domains:

- **Cognitive and metacognitive skills** involve reasoning, critical thinking, problem-solving and the ability to reflect on and regulate one's own learning and professional development.
- **Social and emotional skills** focus on interpersonal abilities such as communication, empathy, collaboration and the regulation of emotions in professional settings.
- **Physical and practical skills** refers to the manual and technical and creative capacities required for tasks. This also includes physical expression that engages the mind and body.

As the demands of the workplace evolve, these skills are increasingly valued for their role in supporting innovation, adaptability, and inclusive practice (OECD, 2019d).

Attitudes and values

Attitudes and values guide behaviour and decision-making. They encompass beliefs, dispositions and ethical orientations that influence how professionals relate to others and the world. In diverse, globalized settings such as THE, values such as respect, fairness, responsibility and integrity are foundational to sustainability and social cohesion (OECD, 2019e). Knowledge, skills, attitudes and values are interdependent and must be developed together to ensure holistic professional growth.

Transformative competencies and the role of agency

The Learning Compass identifies three transformative competencies essential for shaping sustainable futures:

- **Creating new value** is concerned with developing the ability to innovate, generate solutions and contribute constructively to change.
- **Reconciling tensions and dilemmas** refers to the ability to engage with conflicting perspectives and navigate complexity.
- **Taking responsibility** is the ability to act ethically, reflectively and with an awareness of social and environmental impact.

These competencies are supported by **agency**, which refers to an individual's ability to set goals and take purposeful action. In professional development, agency enables proactive learning, career planning and ethical leadership. **Co-agency** further supports this by fostering collaboration and shared responsibility in professional communities (OECD, 2019a).

The Anticipation-Action-Reflection Cycle (AAR) in professional development

The Anticipation-Action-Reflection (AAR) cycle provides a practical tool for embedding the Learning Compass in ongoing development. Professionals begin by anticipating future challenges and opportunities, proceed to take deliberate action, and then engage in critical reflection on outcomes and implications (OECD, 2019f).

In THE settings, the AAR cycle can be used to guide decisions in areas such as guest experience, sustainability policy, and stakeholder engagement. It encourages a structured approach to learning from experience and reinforces the development of transformative competencies over time.

The cycle aligns with Anderson's emphasis on purposeful, reflective practice and supports long-term growth through iterative learning (see Anderson's Five Ps for effective professional development on page 71 for a full discussion of his framework).

ACTIVITY 2: APPLYING THE LEARNING COMPASS – PLANNING YOUR NEXT DEVELOPMENT STAGE

Purpose

To apply your insights from the competency mapping activity (Activity 1) to a realistic professional development plan using the OECD Learning Compass. This task is designed for individual work, but can also be discussed in pairs or small groups to compare perspectives and strategies.

Instructions

1 **Review your competency mapping summary**

  Refer back to your self-assessment from Activity 1. Choose one competency or domain (knowledge, skill, attitude, value or transformative competency) that you identified as a priority for development.

2 **Describe a realistic work or study context**

  Briefly describe a context (current job, placement or project, or future career role) where this competency is particularly relevant.

  o What is the situation or role?

  o Why does this competency matter in that context?

3 **Use the AAR cycle** to structure your development plan

  Using the Anticipation–Action–Reflection cycle, complete the following:

  o Anticipation: What challenges or opportunities do you foresee that require this competency?

> - Action: What specific steps will you take to strengthen it (e.g. seek feedback, engage in a learning activity, practice in role)?
> - Reflection: How will you evaluate your growth? What indicators or outcomes will help you measure progress?
>
> **4 Peer discussion (optional)**
>
> In pairs or small groups, share your development plan. Offer feedback on each other's strategies – what seems effective, realistic or transferable?

Designing effective professional development

As explored throughout this chapter, from developing a sustainability mindset and SDG 4 to grounding sustainability in professional development, UNESCO's key competencies and the OECD Learning Compass, developing sustainability professional competence in THE requires a deliberate and values-driven approach. Professionals must be proactive, reflective and action-oriented, equipped to navigate ethically complex, intercultural and high-pressure environments.

In this context, professional development extends beyond technical training. Richards (2015) stated that professional development serves to articulate and reinforce an organization's ethos. Professional development should be the instrument by which employees make continuous improvement by 'maintaining the interest, creativity, and enthusiasm … in their profession'. Ousseini (2018) further argued that CPD enhances practitioner performance by deepening their understanding of the policies directing their industry. He stated that the point of CPD is to improve the professional's perspective of their work, expand their career opportunities and prepare them for change. Anderson (2018) synthesizes these perspectives through his five Ps of effective professional development framework by recommending an integrated approach where the needs of both the employer and the employee are met.

This is particularly relevant for THE, where issues of staff retention and sustainable capacity-building are urgent. Gabbitas (2023) reported that 30 per cent of employees in hospitality leave their role within the first 90 days. He attributed this attrition to unmet expectations and insufficient training for the job. This highlights that professional development must begin at the point of recruitment and continue throughout employment. Without consistent and structured CPD, THE organizations risk high staff turnover, skills shortages and underprepared staff, all of which will affect progress towards their long-term sustainability goals.

Training for and within the industry must therefore promote the core values associated with global citizenship and sustainable professionalism. CPD offers the mechanism through which individuals can be supported to meet high professional standards, including holding a sustainability mindset.

In the following section, Anderson's five Ps framework is introduced as a practical model for designing professional development. It is followed by a discussion of training needs analysis (TNA), which helps organizations and individuals to identify key areas for development and training and tailor learning interventions accordingly.

Anderson's five Ps of professional development

Anderson's five Ps for effective professional development (PD) emphasize the importance of a unified approach to the training, as illustrated in Figure 3.2. His focus is on the benefit of the training for the employee and the organization. The value to the trainee will also benefit the business because a company will invest in its employees' PD to advantage its own practices. Anderson's five Ps are Purpose, Personalization, Priority, Passion and Professional Learning Community. As Figure 3.2 demonstrates, PD is a continuous process drawing on knowledge, skills and experiences to further professionalism.

Figure 3.2 Anderson's five Ps framework for effective professional development

Anderson, 2018

Purpose

For Anderson (2018), the starting point for an understanding of the purpose of PD for the individual and the organization is that the PD is mutually advantageous. Only then can a clear purpose for the need of the PD be established. This idea challenges the more conventional top-down approach to training.

When thinking about training THE practitioners about sustainability for their industry, the purpose is to cultivate a commitment to sustainable practice. Once the purpose is clear then the individual and the organization can appreciate the individual's current and future contribution to the organization, wider THE industry and the host community.

Anderson's ideas as they apply to the individual and the organization are best illustrated in a hierarchy list, as represented in Figure 3.3. It shows how both stakeholders need to understand the purpose of professional development and its continuation for the mutual benefit of the individual and the organization.

Figure 3.4 is an extract from the full hierarchy list above. It demonstrates how reading down the column presents the development of the individual as separate from the organization as the training progresses.

Reading down the column, the individual is learning about sustainable practices in their work environment. They are learning by trying to solve a problem in a creative and innovative way. As ongoing training becomes part of the individual's work experience, they are recognizing how they are contributing to the standards the industry is required to meet or sets itself in relation to sustainable behaviours. The outcome is that the individual's sustainable behaviours set a standard that promotes a culture of sustainability within the organization.

Figure 3.5 is also an extract from the full hierarchy list. It demonstrates how reading across the tiles in each column demonstrates the relevance of the training to both the individual and the organization.

Reading across the columns, the individual is learning about sustainable practices in their work environment. The benefit of this training to the organization is that they are training an employee to become capable of contributing to the organization's sustainability goals and objectives.

As the individual employee develops their innovative thinking and problem-solving abilities, the organization is benefitting from someone who can work through a process to provide a solution. It is valuable for the individual and the organization to understand the dynamics of purpose as it assigns relevance to the professional development process.

Personalization

The second P in Anderson's framework is Personalization. He suggests asking the question What are my personalized needs for professional development? By asking

Figure 3.3 Applying Anderson's concept of Purpose in the five Ps to professional development in a THE context

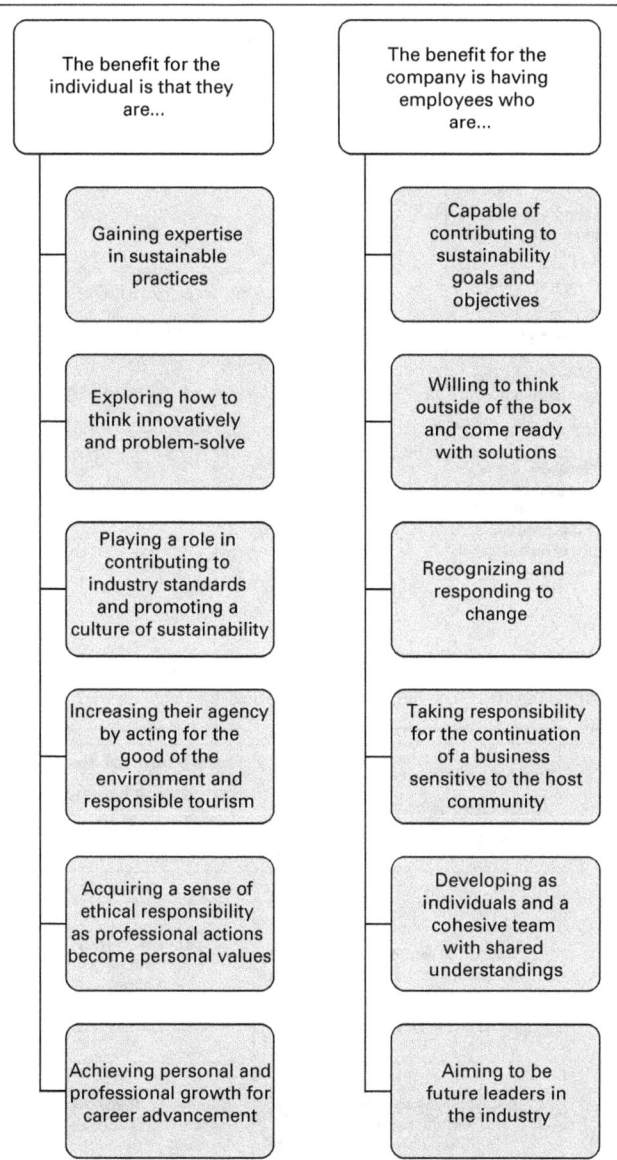

this question, the employee is taking responsibility for their own PD, recognizing that it is central to achieving their goals. It also ensures the PD is relevant and targeted. There is an intention in the process of personalizing PD. For Anderson intentionality is an inherent part of self-improvement and must be built into the PD. Through the intentionality, the employee increases the possibility of learning more about themselves and their capabilities than expected – self-awareness. They may

Figure 3.4 Extract of the application of Purpose to professional development training when reading down a column

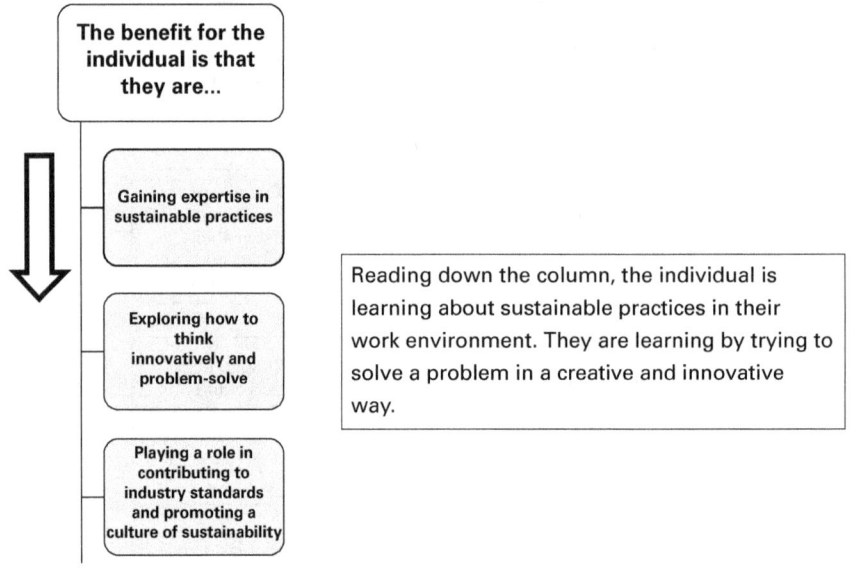

Figure 3.5 Extract of the application of Purpose to professional development training when reading down across two columns

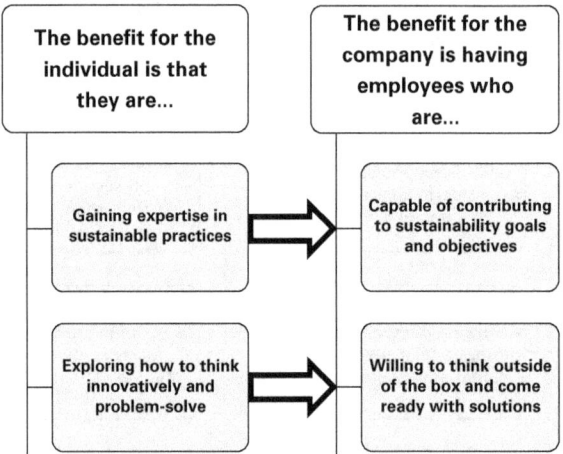

Reading across the columns, the individual is learning about sustainable practices in their work environment. The benefit of this training to the organization is that they are training an employee to become capable of contributing to the organization's sustainability goals and objectives. As the individual employee develops their innovative thinking and problem-solving abilities, the organization is benefitting from someone who can work through a process to provide a solution.

acquire a new understanding about themselves that shapes their professional identity. In this context, the organization can facilitate opportunities for staff to connect personalization with intentionality to identify how to become sustainability citizens.

Priority

For Anderson, personalization cannot be achieved without the third P – Priority. When employees personalize their purpose and identify their intention, they could feel overwhelmed by the many potential areas for self-improvement. This could be especially so when undertaking PD for sustainability. The employee will be trying to balance the needs of the industry, organization, host community, and their own career aspirations. The sense of being overwhelmed can often lead to inaction and lack of confidence. Anderson advises drawing up a list of priorities and rating them in order of importance. In this way the employee can focus on one key area at a time.

Again, the organization and the employee taking an interactive approach to the PD is likely to be more effective. Working together on the question 'What is the one priority that can be identified to work on now?' (Anderson, 2018), there is an agreed outcome to enable meaningful progress.

Passion

When Anderson talks about the fourth P, Passion, he is not only referring to having a passion for the industry but having a passion for PD itself. For Anderson, 'Passion is what drives us to accomplishing our goals' (Anderson, 2018). In other words, purpose, personalization and priority are less likely to be achieved without passion. In the THE industries, a professional's effectiveness and success are significantly influenced by their level of passion. Passion is often described as a deep emotional connection, enthusiasm and dedication to a specific learner activity or goal (Urien and Silas, 2024). It is characterized by a clear sense of purpose, internal motivation and commitment to excellence and personal growth.

For THE professionals, passion involves a genuine desire to seek opportunities for personal growth and to contribute to the success of the organization. Urien and Silas (2024) state that understanding the impact of passion on professional effectiveness can provide valuable insights into the motivational factors that can be addressed in PD to further improve performance.

The implications of nurturing passion through PD involves the co-creation of PD that meets the needs of the individual, the organization and the host community. This is particularly important given Gabbitas' (2003) earlier observation about the high number of people who leave THE within three months. One of the challenges THE will have to address when negotiating PD is a passion, or lack of it, for sustainability. THE employees want to work in THE, not in sustainability. This returns the

PD to considerations of personalization or self-awareness. When a person is highly self-aware, they are more likely to act in ways that are consistent with their personal identity and moral standards, even in group settings. One of the outcomes of PD would be to channel self-awareness for sustainable actions through exploring the moral and ethical implications of not taking a sustainable approach to THE.

Professional Learning Community

Anderson's (2018) fifth and final P is Professional Learning Community. Professional learning communities are groups of professionals who regularly met to plan, solve problems, and learn together (Crandall and Finn Miller, 2014). In industry, learning and growth are often social processes. Engaging with peers and colleagues during PD not only provides support but engenders accountability. Research by Crandall and Finn Miller highlights that when organizations align their performance goals with employee development through professional learning communities, they tend to see positive outcomes. This is because by working together the organization and the employee are creating a culture of learning and development in which all parties can recognize the benefits and reap the rewards. Crandall and Finn Miller, in their work with teachers, suggest enabling them to ask the following questions about their personal responsibility and the collaborative learning process so action can be taken:

> 'How can I take responsibility for my own professional development and create a professional learning community with my trusted colleagues? How can my colleagues and I share what we are learning about our practice as classroom teachers that will continue to improve our teaching and the learning of our students?' (Crandall and Finn Miller, 2014)

Their questions can be reframed to consider the employee THE professional:

1 How can I take charge of my own THE professional development and establish a professional learning community with people I trust, including the host community?
2 How can my colleagues and I share our experiences and insights to continuously improve our practices and sustainability awareness for proactive behaviour change?

Richards (2015) emphasizes that organizations that understand the value of professional development for their staff should adopt a strategic approach to these activities. They should also provide the necessary resources and support to make them effective. By having THE professionals complete a personal PD plan, their responses can be strategically used by the organization to plan professional development initiatives that meet both the individual and organizational needs.

ACTIVITY 3: PLANNING FOR PURPOSE – CONSTRUCTING YOUR PERSONAL PROFESSIONAL DEVELOPMENT PLAN (PPDP)

Purpose

Building on your sustainability competency mapping (Activity 1) and your Learning Compass reflection and planning (Activity 2), this task supports you in creating a Personal Professional Development Plan (PPDP) using Anderson's five Ps framework. This activity encourages you to integrate your insights about your current capabilities, development needs and future aspirations to shape a sustainable, values-driven professional identity.

This activity is designed for individual reflection but may be discussed in pairs or small groups to deepen insight and generate per feedback.

Instructions

1 **Review your previous activities**

Look back at the competencies you rated and prioritized for development (Activity 1) and the professional context and AAR plan you drafted (Activity 2). These will form the foundation of your personal development planning.

2 **Use Anderson's five Ps** to guide your thinking

The 5Ps	Questions to consider	Your thoughts	How does this link to sustainability in THE?
Purpose	What is the purpose of my PD as a current or future THE professional?		
Personalization	Based upon my earlier reflections, what are my personalized needs and areas for growth in PD?		
Priority	Of all the areas identified, what is the one priority I will work on now, and why?		
Passion	What motivates me to pursue this goal, and how will I maintain my passion to achieve my goal?		
Professional Learning Community	Who can support me in this journey (mentors, peers, networks) and how can I contribute in return?		

> **3 Next step**
> Once you complete the table, review your responses and identify a concrete action you will take in the next 30 days to begin moving towards your goal. This might be enrolling in a course, scheduling a conversation with a mentor or starting a small-scale sustainability project.

> **REFLECTIVE QUESTION**
>
> How does aligning your personal development with sustainability values change the way you think about your future career in THE?

Using Anderson's five Ps to construct a personal development plan for a career in THE is a useful way to understand your needs in a professional environment. However, professional development cannot be achieved in isolation; it has to consider the wider context in which it is to be employed.

Training needs analysis for sustainability in THE

There is a substantial amount of literature discussing training needs analysis (TNA) and its variants, which will not be addressed here. According to the recruitment agency Indeed.com (.n.d.), effective CPD can only be designed if a TNA is undertaken. TNA is a method to assess the training required by an organization of its employees. It identifies gaps in knowledge and skills to determine the necessary training for employees to perform their roles effectively. Alzahmi and Alshamsi (2024) identify four types of training organizations undertake:

1 **Induction training** – the initial orientation and training provided to new employees to familiarize them with the organization's policies, procedures, culture and job responsibilities.
2 **Job training** – the process of training employees in the specific skills and knowledge required to perform their jobs effectively.
3 **Training for promotion** – the training required to prepare employees with the advanced skills and knowledge necessary for qualifying for a higher-level position in an organization.
4 **Refresher training** – the training to reinforce employees' existing knowledge and skills as well as update them to ensure the employee remains effective and current in their role.

For any of these types of training to be designed and effectively implemented, conducting a TNA is the initial step in the training process (Alzahmi and Alshamsi, 2024). Once a knowledge gap is identified, the specific training need can be pinpointed. The analysis can focus on a particular team or department or encompass the entire organization. The TNA process begins by examining the overall organizational goals and considering the skills employees need to achieve these goals. It then explores the specific types of training required for the PD.

Alzahmi and Alshamsi highlight the importance of creating a culture of continuous learning in an organization through TNA. They identify a three-tiered approach to TNA when considering training goals, as illustrated in Figure 3.6.

By employing the three-tiered approach, the short and long-term goals of the organization and the training needs of its employees can be ascertained. Once the knowledge and skills gaps of the organization and the employees are understood, then CPD can designed at the organizational, operational and personal level.

The top-down approach to professional development

The top-down approach is a structured, hierarchical method taken by those in leadership positions to drive the development initiatives required for the business. This includes taking a top-down approach to PD. When adopting a top-down approach the organization will align training to its strategic goals and objectives to ensure that its employees' development supports the broader business vision. There are several advantages to this approach, especially when considering the induction training of newly appointed employees.

Figure 3.6 Alzahmi and Alshamsi's three-tiered approach to training needs analysis

Organizational Analysis
- assesses the focus areas of training
- identifies the organization's goals
- evaluates how effectively training enhances performance
- highlights development needs in specific departments

Operational Analysis
- identifies knowledge, skills and abilities needed to perform specific tasks
- informed by data sources such as job descriptions, quality assurance procedures and performance standards
- may use interviews with line management to explore operational needs

Personal Analysis
- evaluates the knowledge and skills of individuals or teams
- identifies training needs by comparing current performance with expected performance
- uses methods such as interviews, questions and skill assessments

Employees new to an organization within a complex industry such as THE are unlikely to know what their professional development needs are. In top-down training, employees receive guidance on what skills and knowledge are important for their roles and for the organization. In relation to grounding sustainability, the organizational sustainability culture can be communicated through the training. This direction helps the employee focus their efforts on developing competencies that are most valued by the organization, and most needed by the industry. As Figure 3.7 illustrates, a top-down approach places the responsibility for sustainability training with the industry and its organizations.

However, it is important to consider the role of the new employee here. The key question to ask in relation to CPD is: 'Who is preparing the professional development opportunity?' The top-down approach is an 'expert driven process' (Richards, 2015) in which the organization is the main source of 'expert' information and knowledge. The role of the employee is to be a passive recipient of workplace training. In this scenario, it is the organization's needs that are being met without consideration of the needs of the employee.

The bottom-up approach to professional development

The bottom-up approach to PD is centred around the idea that it is the individuals, rather than the leadership or organization, to take primary responsibility for their own growth and learning. The PD is driven by the individual employee who decides

Figure 3.7 Top-down approach for THE to shift to a sustainable tourism mindset

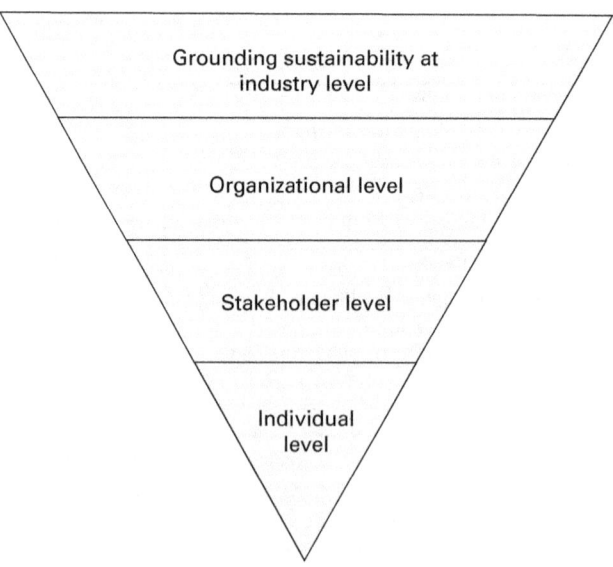

Figure 3.8 Bottom-up approach whereby the individual initiates PD for a sustainable tourism mindset

how they want to approach their own development. The bottom-up approach emphasizes personal agency and the belief that employees are best positioned to understand their own developmental needs.

This approach is most effectively used with experienced, motivated individuals who can hold themselves accountable for their professional growth as shown in Figure 3.8. For these employees, the bottom-up approach could lead to the design of a personal CPD plan targeting a long-term employment goal. The bottom-up model would not be so effective with new employees who hold a limited understanding of the organization's ethos concerning sustainability and the opportunities open to them.

REAL-WORLD EXAMPLE   Green Pearls®

The communication and information platform for unique, sustainable places for tourism Green Pearls® shares its vision to 'Do good and talk about it'. Read about Green Pearls® and then complete Activity 4.

Green Pearls® is an international platform dedicated to promoting sustainable tourism by showcasing eco-friendly hotels and destinations that prioritize environmental responsibility, cultural preservation and social equity. Founded on the principle of 'Do good and talk about it,' Green Pearls® collaborates with accommodations that adhere to

Figure 3.9 Green Pearls®

high sustainability standards. By sharing these success stories, they inspire both travellers and businesses to adopt responsible tourism practices.

Sustainability approach

Green Pearls® evaluates and supports member establishments based on a comprehensive set of sustainability criteria, ensuring that at least 80 per cent of these standards are met. Their approach includes:

- **Management plan**: The business has established a long-term sustainability management system tailored to its scale and operational needs, incorporating environmental, quality, health and safety considerations. The hotel fully complies with all relevant local, national and international regulations, including those related to health, safety, environmental protection and labour standards.
- **Eco-friendly infrastructure**: Green Pearls® members incorporate sustainable designs using local materials, energy-efficient technologies and green spaces to minimize environmental impact.
- **Resource conservation**: Hotels implement measures to reduce water and energy consumption through innovative solutions such as solar power, rainwater harvesting and waste reduction programmes.
- **Biodiversity protection**: Accommodations take steps to preserve local flora and fauna, integrating nature conservation efforts into their operations.
- **Community and employee engagement**: Properties focus on fair labour practices, local employment and community-based initiatives to ensure tourism benefits extend beyond guests to the surrounding population.
- **Sustainable gastronomy**: Many establishments source food locally, favouring organic and fair trade products to support regional farmers and reduce carbon footprints.
- **Cultural commitment**: Hotels actively engage in preserving local traditions and heritage, offering authentic experiences while respecting Indigenous communities.

Through these efforts, Green Pearls® has created a model for sustainable tourism that aligns profitability with environmental and social responsibility (Green Pearls, .n.d.).

ACTIVITY 4: DESIGNING A SUSTAINABILITY TRAINING PLAN USING TNA

Purpose

This activity shifts your focus from personal development to organizational training. You will design a sustainability-oriented professional training plan for a department or role within a THE organization using the structure of training needs analysis (TNA). Your plan should reflect an understanding of how sustainability values can be embedded at all levels of professional practice. To support your thinking, refer to sustainability practice demonstrated by Green Pearls® properties, which integrate community involvement, resource efficiency and ethical business into their training and operations.

Instructions

1. **Select a department or role**

   Choose a professional role or department in THE that would benefit from sustainability training. Options include:

   - hotel management
   - housekeeping
   - food and beverage service
   - front desk and guest services
   - events and conference planning
   - cultural experience coordination or interpretation

2. **Identify key training areas**

   From the list below, select TWO areas your training plan will address:

   - Sustainable operations: Educate staff on energy conservation, waste reduction and water management techniques.
   - Guest engagement: Train employees on how to inform and encourage guests to participate in sustainable practices.
   - Local community involvement: Provide strategies for staff to support community projects and cultural preservation efforts.
   - Ethical business practices: Emphasize fair labour policies, workplace inclusivity and corporate social responsibility.
   - Eco-friendly hospitality practices: Train housekeeping and food service teams on sustainable products and sourcing.
   - Crisis management: Prepare staff for environmental challenges, such as natural disasters or resource shortages, in a sustainable manner.

3 **Conduct a training needs analysis**
   Use the table below to identify sustainability training needs at the organizational, operational and individual levels. Consider how each level supports the integration of sustainability into the workplace practice. Think about how the values seen in the Green Pearls® examples can inform your content.

TNA level	Key questions	Your analysis
Organizational	What sustainability goals or brand commitments does the organization have? How can these be reflected in the training?	
Operational	What procedures or team behaviours need to change or improve to align with sustainable operations?	
Individual	What specific sustainability related competencies, knowledge or attitudes do employees need to develop?	

4 **Training design brief**
   Based on your TNA table, draft a training plan that includes:
   o target participants (who will be trained and why)
   o learning outcomes (what trainees will know/do after training)
   o training format (e.g. workshop, coaching, peer-led sessions, digital module)
   o connection to Green Pearls® values
   o evaluation or follow-up methods to assess training impact

   Your brief can be written as a short report, slide presentation or one-page training summary.

5 **Peer discussion (optional)**
   Pair with a classmate and exchange training briefs. Compare how different departments require different sustainability competencies and discuss how your TNA influenced your design choices.

Conclusion

As the global landscape continues to change environmentally, socially and economically, professionals entering THE industries need to take responsibility for how they contribute to sustainable futures. The learning frameworks and tools presented in this chapter offer both structure and purpose to the ongoing process of professional development.

For early career professionals the key challenge is not simply acquiring knowledge but learning how to apply values, think critically and act with integrity in changing and often uncertain contexts. Building a professional identity rooted in sustainability means learning how to navigate tension. Lead change and cultivate a mindset that balances personal purpose with collective responsibility.

In moving forward, professional development should not be treated as a one-off requirements or training event. It should be experienced as a career-long commitment to learning, reflection and collaboration in service of both personal growth and global goals.

KEY TAKEAWAYS

- Professional development is a shared responsibility:
  - Both individuals and organizations must commit to ongoing learning that aligns with sustainability values.
- Sustainability competencies are practical and learnable:
  - Skills such as systems thinking, ethical reasoning and collaboration are essentials for all roles in THE.
- Mindset influences professional success:
  - A growth-oriented, reflective mindset is crucial for responding to uncertainty and leading change.
- Agency and co-agency drive impact:
  - Professionals must take ownership of their learning, while also supporting and learning from others.
- Effective training needs structure and purpose:
  - Tools such as TNA and the five Ps help align professional development needs with meaningful organizational outcomes.

REFLECTIVE QUESTIONS

1. What is meant by having a sustainability mindset? How has your mindset influenced the way you approach sustainability in everyday situations?
2. Which UNESCO competency do you think will be the most challenging to develop, and why?
3. How can you use the five Ps framework to prepare a personalized continuous professional development plan?
4. What does a meaningful professional development plan for sustainability competence look like for you in your first year of employment?

References

Alzahmi, A and Alshamsi, S (2024) The influence of applying human resource training need analysis on employee's performance, *Journal of Human Resource and Leadership*, 9, pp 1–18, https://doi.org/10.47604/jhrl.2302 (archived at https://perma.cc/W57Q-CPZ5)

Anderson, N J (2018) The five Ps of effective professional development for language teachers, *MEXTESOL Journal*, **42** (2), pp 1–9, https://mextesol.net/journal/index.php?page=journal&id_article=3416 (archived at https://perma.cc/K3K7-P53U)

CIPD (.n.d.) About CPD, www.cipd.org/uk/learning/cpd/about/ (archived at https://perma.cc/HGJ2-XX55)

Crandall, J and Finn Miller, S (2014) Effective professional development for language teachers. In M Celce-Murcia, D M Brinton and M A Snow (eds.) *Teaching English as a second or foreign language*, National Geographic Learning/Cengage Learning, Boston, MA

Dweck, C (2015) Teaching a Growth Mindset, YouTube, 3 November, www.youtube.com/watch?v=isHM1rEd3GE (archived at https://perma.cc/779W-WLMD)

EduSkills OECD (2019) OECD Future of Education and Skills 2023: OECD Learning Compass 2030, YouTube, 17 May, www.youtube.com/watch?v=M3u1AL_aZjI (archived at https://perma.cc/6CCZ-D3VS)

Fernández, J E (2021) Grounding sustainability: Strategies and opportunities for industry [webinar], MIT Professional Education, 20 May, https://professionalprogrammes.mit.edu/grounding-sustainability-webinar (archived at https://perma.cc/C7YW-NRJG)

Gabbitas, B (2023) How the hospitality industry can prioritise retention, People Management, 22 March, www.peoplemanagement.co.uk/article/1817348/hospitality-industry-prioritise-retention (archived at https://perma.cc/MY93-FN6G)

Green Pearls (.n.d.) Meet Us – What we believe and how to become a member, www.greenpearls.com/meet-us/ (archived at https://perma.cc/6ZGR-CCWD)

Indeed.com (.n.d.) What is training needs analysis and why is it useful? Indeed Career Guide, https://uk.indeed.com/career-advice/career-development/training-needs-analysis

Manpower Group (2024) A people-first green business transformation, 5 January, www.manpowergroup.com/en/insights/report/a-people-first-green-business-transformation (archived at https://perma.cc/QG2H-MPHX)

OECD (2019a) Learning Compass for 2030: Conceptual learning framework, www.oecd.org/content/dam/oecd/en/about/projects/edu/education-2040/1-1-learning-compass/OECD_Learning_Compass_2030_Concept_Note_Series.pdf (archived at https://perma.cc/XEP7-XS8W)

OECD (2019b) Transformative Competencies for 2030 concept note, www.oecd.org/content/dam/oecd/en/about/projects/edu/education-2040/concept-notes/Transformative_Competencies_for_2030_concept_note.pdf (archived at https://perma.cc/NP8S-77FV)

OECD (2019c) Knowledge for 2030 concept note, www.oecd.org/content/dam/oecd/en/about/projects/edu/education-2040/concept-notes/Knowledge_for_2030_concept_note.pdf (archived at https://perma.cc/F4BW-XB6B)

OECD (2019d) Skills for 2030 concept note, www.oecd.org/content/dam/oecd/en/about/projects/edu/education-2040/concept-notes/Skills_for_2030_concept_note.pdf (archived at https://perma.cc/8BBN-5F3P)

OECD (2019e) Attitudes and Values for 2030 concept note, www.oecd.org/content/dam/oecd/en/about/projects/edu/education-2040/concept-notes/AAR_Cycle_concept_note.pdf (archived at https://perma.cc/EB98-KMQS)

OECD (2019f) Anticipation-Action-Reflection Cycle for 2030 concept note, www.oecd.org/content/dam/oecd/en/about/projects/edu/education-2040/concept-notes/AAR_Cycle_concept_note.pdf (archived at https://perma.cc/EB98-KMQS)

Ousseini, H (2018) PD for NEST/NNEST ELT teachers in the EFL setting. In *TESOL Encyclopaedia of English Language Teaching, Teacher Training and Professional Development*, Wiley & Sons, www.researchgate.net/publication/373241629_PD_for_NEST_NNEST_ELT_Teachers_in_the_EFL_Setting (archived at https://perma.cc/6N3W-BF7E)

Richards, J C (2015) *Key Issues in Language Teaching*, Cambridge University Press, Cambridge

Rimanoczy, I (2020) *The Sustainability Mindset Principles: A guide to developing a mindset for a better world*, Routledge, Abingdon

RTVMalacanang (2024) Opening Ceremony of the 36th CAP-CSA, 28 June, www.youtube.com/watch?v=O4cPr_wGngc (archived at https://perma.cc/LW8U-JEAB)

Sustainability Stories (2023) How do you start a green team at a hotel? with Dr Aurora Dawn Benton, series 1, episode 25, 30 January, https://sustainabilitykiosk.com/podcast/s1e25-how-do-you-start-a-green-team-at-a-hotel-explained-by-dr-aurora-dawn-benton-founder-chief-change-agent-at-astrapto/ (archived at https://perma.cc/L98A-JXJ4)

UN (.n.d.) Goal 4, https://sdgs.un.org/goals/goal4 (archived at https://perma.cc/G5F7-7QAR)

UNESCO (2017) Education for Sustainable Development: Learning objectives, https://unesdoc.unesco.org/ark:/48223/pf0000247444 (archived at https://perma.cc/VF66-AYNN)

Urien, J and Silas, C (2024) Passion for teaching and teachers' teaching effectiveness in secondary schools in delta state, *Social Science and Humanities Journal*, 8 (7), pp 4188–96, https://sshjournal.com/index.php/sshj/article/view/1170 (archived at https://perma.cc/CF4D-8SQ5)

4 | Responsible practice for cultural sustainability

| Anticipatory competency | Critical thinking competency | Self-awareness competency | Normative competency |

CHAPTER AIM

This chapter encourages a practical understanding of how tourism, hospitality and events (THE) industries can simultaneously support and damage the sustainability of a community's cultural practices. By exploring the complexities of sustainable cultural tourism and the role THE should play, the chapter considers the challenges faced when collaborating with host communities to avoid ethnocentrism, misrepresentation and cultural appropriation. It integrates corporate social responsibility (CSR) and stakeholder theory as interconnected and complementary frameworks to address the dual objectives of sustainability and business growth. The chapter also explores ways of guiding thinking, emotional engagement and behavioural actions towards achieving social and cultural sustainability in a professional THE context.

This chapter explores the complexities of cultural sustainability within THE industries. It examines how THE practices can either contribute to the safeguarding of cultural heritage or perpetuate harm through misrepresentation, over-commodification and exclusion. Drawing on UNESCO's cultural conventions directives and SDG 8, SDG 11 and SDG 12, the chapter encourages the development of ethically grounded and community-centred approaches to cultural tourism. Through critical case studies, reflective activities and stakeholder analysis, learners will build key UNESCO sustainability competencies – including critical thinking, self-awareness, anticipatory thinking, and normative reasoning – equipping them to support meaningful cultural engagement and long-term sustainability in THE contexts.

LEARNING OUTCOMES

Upon completion of this chapter, you will be able to:

- understand the role THE should play in promoting social and cultural sustainability
- recognize the importance of safeguarding cultural identity and heritage as characteristics of human rights and dignity
- understand the interrelatedness of types of cultural heritage
- identify THE business practices and assumptions that support and damage the sustainability of a community's cultural heritage
- explain how corporate social responsibility and stakeholder theory have a role to play in THE's contribution to social sustainability
- describe the role of cultural sustainability as it contributes to the SDGs

KEY WORDS

Cultural diversity, cultural sustainability, tangible and intangible heritage, overtourism, social sustainability, corporate social responsibility, stakeholders.

Introduction

Tourism has been defined by UN Tourism (.n.d.) as a social, cultural and economic phenomenon. Within this context, culture refers to a collection of spiritual, material, intellectual and emotional characteristics that distinguish a nation. These characteristics encompass lifestyle, value systems, traditions and beliefs. According to UNESCO, these dimensions of culture are fundamental to the identity of individuals and communities, and their preservation is essential to safeguarding human dignity (UNESCO, 2001; UNESCO, 2005).

Culture has been recognized as one of the four pillars of sustainability, alongside the environmental, social and economic dimensions. Its role in sustaining the beliefs, practices and heritage of communities is considered vital to the continued development of societies. In response to the increasing pressures placed on culture by globalization and tourism, UNESCO has developed several frameworks to position culture as central to sustainable development:

- The Universal Declaration of Cultural Diversity (2001) affirms cultural diversity as a common heritage of humanity and links it to human rights and dignity.
- The Convention for the Safeguarding of the Intangible Cultural Heritage (2003) identifies living practices as essential to cultural continuity.
- The Convention on the Protection and Promotion of the Diversity of Cultural Expressions (2005) promotes policy measures to support cultural expression and creative industries.

This chapter explores the intersection between these cultural sustainability frameworks and THE industries. It focuses on the concept of sustainable cultural tourism and considers how THE can contribute positively to the safeguarding of cultural heritage. At the same time, it recognizes the risks that tourism poses to cultural practices when activities are driven by external market interests or developed without meaningful community engagement.

The analysis in this chapter is structured around two key questions that reflect the professional and ethical responsibilities of those working in THE, as identified by Butvin (2023):

1 How can tourism experiences support the sustainability of a community's most cherished cultural practices?
2 How can tourism experiences damage the sustainability of a community's most cherished cultural practices?

These questions frame the critical issues examined throughout the chapter. They provide a lens through which to evaluate the effects of tourism development on cultural heritage, and they guide consideration of how tourism experiences can be designed or adapted to promote cultural sustainability. By foregrounding these questions, the chapter emphasizes that cultural heritage is not a neutral backdrop for tourism, but a living system shaped by social values, historical knowledge and ongoing community participation.

To explore these issues, the chapter first defines cultural heritage and introduces key terms used in the study of sustainable cultural tourism. It then examines the challenges to heritage posed by practices such as overtourism, cultural appropriation and over-commodification. These challenges are analyzed in relation to international cultural policy, professional ethics and industry frameworks. The chapter concludes by presenting strategies to support sustainable tourism practice, including the application of corporate social responsibility, stakeholder theory and the UN Sustainable Development Goals.

Sustainable cultural tourism depends on the ability of THE professionals to recognize cultural heritage as a dynamic and community-rooted system. This requires a commitment to inclusive, transparent and respectful engagement with the communities whose heritage is shaped by and shared through tourism experiences.

Understanding cultural heritage

The concept of cultural heritage encompasses the material and immaterial elements of culture that communities recognize as meaningful to their identity, continuity and way of life. Cultural heritage connects people to place and memory, and its value extends beyond historical preservation to include contemporary relevance and community wellbeing. In the context of THE, cultural heritage shapes destination identity and plays a central role in the visitor experience. Its significance is not only aesthetic or economic but also ethical and relational, grounded in the responsibilities of THE professionals to engage with cultural practices in ways that respect their origins and social meaning.

Cultural heritage is commonly understood in two broad forms: tangible and intangible. Tangible cultural heritage includes physical artefacts and sites that hold historical, artistic, scientific or symbolic significance. This may encompass monuments, architectural structures, landscapes, and collections held in museums. Intangible cultural heritage refers to the living practices, expressions, knowledge systems and skills that are transmitted across generations and maintained through active participation. These may include languages, oral histories, social customs, music, ritual performance, traditional craftsmanship and other forms of cultural expression.

UNESCO has articulated a comprehensive understanding of cultural heritage across its declarations and conventions. The 2003 Convention for the Safeguarding of the Intangible Cultural Heritage defines intangible heritage as dynamic and responsive to environmental and social change. It stresses that intangible heritage must be recognized and sustained by the communities who practice it, and that its value lies in the sense of identity and belonging it provides. This understanding is supported by the 2001 Universal Declaration on Cultural Diversity, which positions cultural diversity as essential to humanity and calls for international cooperation in its protection. The 2005 Convention on the Protection and Promotion of the Diversity of Cultural Expressions expands this view by recognizing the need to protect the cultural and creative industries that support local expression and participation in the global cultural economy.

From a THE perspective, acknowledging the significance of cultural heritage to tourist destinations, as well as the sector's responsibility for preserving and conserving that heritage, is fundamental to achieving sustainable cultural tourism. Cultural heritage should not be treated as a static attraction or a consumable product. Instead, it should be understood as a set of community-based practices that are continuously reshaped by those who live them. The ability of tourism experiences to support or damage these practices depends on how they are framed, delivered and governed.

Two central questions posed by Butvin (2023) frame the evaluation of cultural engagement in tourism contexts:

1 How can tourism experiences support the sustainability of a community's most cherished cultural practices?

2 How can tourism experiences contribute to the weakening or distortion of those same practices?

These questions highlight the importance of professional judgement, community collaboration and reflective practice in the design and management of tourism experiences. They also highlight the need for cultural heritage to be situated within broader conversations about power, representation and equity in the tourism industry. Cultural heritage can be understood in terms of two key elements: people and place. In the context of THE, cultural heritage encompasses both tangible and intangible forms.

To support a consistent understanding of the concepts relevant to cultural heritage, the following key terms are used:

Heritage

'Heritage is our legacy from the past, what we live with today, and what we pass on to future generations. Our cultural and natural heritage are both irreplaceable sources of life and inspiration.' (UNESCO, .n.d.)

Natural heritage

'Natural heritage refers to natural features, geological and physiographical formations and delineated areas that constitute the habitat of threatened species of animals and plants and natural sites of values from the point of view of science, conservation or natural beauty. It includes private and publically [sic] protected natural areas, zoos, aquaria and botanical gardens, natural habitat, marine ecosystems, sanctuaries, reservoirs etc.' (UNESCO Institute for Statistics, 2009)

Cultural heritage

'Cultural heritage includes artefacts, monuments, a group of buildings and sites, museums that have a diversity of values including symbolic, historic, artistic, aesthetic, ethnological or anthropological, scientific and social significance. It includes tangible heritage (movable, immobile and underwater), intangible cultural heritage (ICH) embedded into cultural, and natural heritage artefacts, sites or monuments. The definition excludes ICH related to other cultural domains such as festivals, celebration etc. It covers industrial heritage and cave paintings.' (UNESCO Institute for Statistics, 2009)

Cultural diversity

'Many ways in which the different cultures of groups and societies find expression. These cultural expressions are passed on within and among groups and societies, and from generation to generation. Cultural diversity, however, is evident not only in the varied ways in which cultural heritage is expressed, augmented and transmitted but also in the different modes of artistic creation, production, dissemination, distribution and enjoyment, whatever the means and technologies that are used.' (UNESCO, 2005)

Intangible cultural heritage

'Practices, representations, expressions, knowledge, skills – as well as the instruments, objects, artefacts and cultural spaces associated therewith – that communities, groups and, in some cases, individuals recognize as part of their cultural heritage. This intangible cultural heritage, transmitted from generation to generation, is constantly recreated by communities and groups in response to their environment, their interaction with nature and their history, and provides them with a sense of identity and continuity, thus promoting respect for cultural diversity and human creativity.' (UNESCO, 2003)

Traditional knowledge

'Knowledge, innovations and practices of indigenous and local communities around the world. Developed from experience gained over the centuries and adapted to the local culture and environment, traditional knowledge is transmitted orally from generation to generation. It tends to be collectively owned and takes the form of stories, songs, folklore, proverbs, cultural values, beliefs, rituals, community laws, local language and agricultural practices, including the development of plant species and animal breeds. Traditional knowledge is mainly of a practical nature, particularly in such fields as agriculture, fisheries, health, horticulture, forestry and environmental management in general.' (Convention on Biological Diversity, .n.d.)

The most appealing cultural heritage destinations offer a combination of both elements (Jones, 2023). According to Butvin (2023) and Jones (2023), a successful destination has a unique sense of place, which is reflected in the landscape and built environment, as well as opportunities for visitors to engage with people who actively practice and sustain cultural traditions. The presence of living cultural heritage should be evident throughout the destination (Butvin, 2023). Butvin further emphasizes that when presenting cultural heritage to visitors, it is important to highlight the connection between the past and the present. While displays of past traditions have educational value, travellers increasingly seek engagement with culture as it is lived and practiced today.

Several examples illustrate how intangible cultural heritage functions in practice.

The Japanese tea ceremony, known as *chanoyu*, reflects the evolution of a tradition rooted in Zen Buddhist values such as harmony, respect, purity and tranquillity. While its core principles remain, contemporary interpretations often incorporate new aesthetics and tools that reflect changing cultural contexts (Pajon, 2024). This example demonstrates how intangible heritage is both historical and adaptive.

The Mediterranean diet, recognized by UNESCO as a form of shared intangible heritage, is another example (see Figure 4.1). Practiced across multiple countries, including Greece, Italy, Morocco and Spain, it represents common culinary traditions that connect communities through seasonal food, social rituals and ecological knowledge. While it may develop cultural bonds, it also raises questions of ownership and identity when nations assert exclusive claims over particular elements of the tradition.

Flamenco, a music and dance tradition from Andalusia in Spain, further illustrates how cultural heritage is deeply embedded in community life. Its value lies not only in the performance but in the lived experience of practitioners who sustain it through local knowledge, emotion and identity. As a form of cultural tourism, flamenco presents opportunities for economic benefit but also carries risks of distortion when removed from its social context and commodified for visitor consumption.

Similarly, the Kente weaving tradition in Ghana reflects a strong connection between culture, identity and place. Woven by Ashanti and Ewe communities, Kente cloth holds symbolic meanings tied to social status, history and community pride. While it has become a popular cultural item in global markets, efforts by local communities to maintain control over its design and usage highlight the importance of community agency in managing cultural expressions (Our Homeland Ghana, 2023).

When considering the connection between sustainability and cultural tourism, the definition offered by the European Commission and the European Cultural Tourism Network (ECTN) is useful. The ECTN accepts the European Commission's definition of sustainable cultural tourism as 'the integrated management of cultural heritage and tourism activities in conjunction with the local community, creating social, environmental, and economic benefits for all stakeholders in order to achieve tangible and intangible cultural heritage conservation and sustainable tourism development' (ECTN, .n.d.). This

Figure 4.1 The adoption of the Mediterranean diet on the Representative List of the Intangible Cultural Heritage of Humanity

Decision of the Intergovernmental Committee: 8.COM 8.10

The Committee

1   Takes note that Cyprus, Croatia, Spain, Greece, Italy, Morocco and Portugal have nominated **Mediterranean diet** (No. 00884) for inscription on the Representative List of the Intangible Cultural
Heritage of Humanity:
   The Mediterranean diet involves a set of skills, knowledge, rituals, symbols and traditions concerning crops, harvesting, fishing, animal husbandry, conservation, processing, cooking, and particularly the sharing and consumption of food. Eating together is the foundation of the cultural identity and continuity of communities throughout the Mediterranean basin. It is a moment of social exchange and communication, an affirmation and renewal of family, group or community identity. The Mediterranean diet emphasizes values of hospitality, neighbourliness, intercultural dialogue and creativity, and a way of life guided by respect for diversity. It plays a vital role in cultural spaces, festivals and celebrations, bringing together people of all ages, conditions and social classes. It includes the craftsmanship and production of traditional receptacles for the transport, preservation and consumption of food, including ceramic plates and glasses. Women play an important role in transmitting knowledge of the Mediterranean diet: they safeguard its techniques, respect seasonal rhythms and festive events, and transmit the values of the element to new generations. Markets also play a key role as spaces for cultivating and transmitting the Mediterranean diet during the daily practice of exchange, agreement and mutual respect.

1   Decides that, from the information included in the file, the nomination satisfies the following criteria for inscription on the Representative List:

UNESCO, 2013

definition highlights the importance of balancing heritage conservation with local participation and benefit-sharing, providing a practical framework for applying cultural sustainability principles in tourism contexts.

In a career within THE industries, understanding and respecting both tangible and intangible cultural heritage is crucial. Sustainable tourism requires operations to actively safeguard both forms of heritage. Butvin (2023) emphasizes the importance of meaningful engagement with host communities to ensure that THE practices respect the authenticity and integrity of intangible cultural heritage. She also points to the critical role of women in preserving and transmitting cultural practices, which is a key consideration in the industry. Collaboration with communities is essential for gaining valuable insights into how to navigate cultural heritage responsibly and effectively. This approach supports inclusive tourism that values cultural continuity, representation and self-determination.

> **ACTIVITY 1: UNDERSTANDING CULTURAL HERITAGE**
>
> *Instructions*
>
> Using the explanation of key terms, place each example of cultural and natural heritage next to its category. One has been done for you.
>
> A. The Parthenon, Greece
> B. The Angkor Temple, Cambodia
> C. The Rice Terraces of the Philippine Cordilleras
> D. ~~The Moai Statues, Easter Island, Chile~~
> E. The Alhambra, Spain
> F. Machu Picchu, Peru
> G. Serengeti National Park, Tanzania
> H. Punakaiki Marine Reserve, New Zealand
> I. Mount Vesuvius, Italy
> J. Yellowstone National Park, USA
> K. Victoria Falls, Zimbabwe, South Africa
> L. Cappadocia Cave Dwellings, Türkiye
> M. Sacred Aboriginal site of Uluru, Australia
> N. Swiss Alps, Switzerland
> O. Galápagos Islands, Ecuador
>
> **Monuments**
> D. The Moai Statues, Easter Island, Chile
>
> **Groups of buildings**
>
> **Sites**
>
> **Natural features**
>
> **Geographical formations**
>
> **Natural sites**

International frameworks for cultural sustainability

Efforts to promote and protect cultural heritage within tourism, hospitality and events are shaped by a set of international frameworks developed by UNESCO. These frameworks provide legal, ethical and policy foundations for supporting cultural sustainability in the face of global change. They affirm the value of cultural diversity, recognize the dynamic nature of intangible heritage and promote international cooperation in the creation, exchange and safeguarding of cultural expressions. Although designed primarily for adoption by national governments, these instruments also offer clear principles that inform professional practice within THE industries.

UNESCO Universal Declaration on Cultural Diversity (2001)

The Declaration positions cultural diversity as a shared resource for humanity. It defines diversity not only as a source of creativity and innovation but also as essential to dialogue, social cohesion and human dignity. The Declaration affirms that the right to maintain and express one's culture is a human right and stresses the importance of cultural identity in the development of inclusive societies. It calls for policies that promote cultural expression and protect cultural goods and services from being reduced solely to commercial products. By identifying culture as both a value and a resource, the Declaration situates cultural sustainability as an integral part of broader development strategies.

UNESCO Convention for the Safeguarding of the Intangible Cultural Heritage (2003)

The Convention builds upon the Declaration by establishing a legal and operational framework for the protection of living heritage. It defines intangible cultural heritage as the practices, representations, knowledge and skills that communities recognize as part of their cultural legacy. The Convention underscores the interconnectedness of intangible and tangible heritage and natural environments. It identifies the threats posed by globalization and over-commercialization and stresses the importance of involving communities – particularly Indigenous and local groups – in safeguarding their own traditions. The Convention also emphasizes the importance of education in transmitting intangible cultural heritage to future generations and advocates for cooperative safeguarding practices that respect the rights and knowledge systems of communities.

UNESCO Convention on the Protection and Promotion of the Diversity of Cultural Expressions (2005)

The 2005 Convention expands the focus of the earlier 2001 Declaration and 2003 Convention by addressing the structural conditions needed to support cultural diversity in the context of globalization. It affirms the right of states to formulate and implement policies that protect local cultural production and promote access to diverse cultural content. The Convention calls attention to the unequal distribution of cultural resources and the risk of market dominance by a limited number of cultural industries. It promotes international collaboration, particularly between developed and developing countries, and supports the development of local cultural and creative industries. The Convention links cultural expression to economic development and recognizes the potential of culture-based sectors to contribute to inclusive and sustainable growth.

Together, these frameworks articulate a comprehensive approach to cultural sustainability. They position cultural heritage as a living and evolving system, rooted in the practices of communities and supported by policy environments that value equity, participation and cooperation. Within THE, these principles can guide the development of experiences that respect cultural integrity and promote meaningful engagement between visitors and host communities.

Professional application of these frameworks involves more than regulatory compliance. It requires a values-based approach that recognizes communities as the primary stewards of their heritage. In this context, cultural heritage is not a passive asset to be consumed but an active domain of meaning and identity that must be protected through respectful and inclusive practices. The frameworks emphasize that sustainable cultural tourism depends on partnerships that centre community voices and ensure that tourism activities support, rather than displace, the cultural practices that define a place.

The role of THE professionals includes interpreting these international principles in relation to local contexts. This involves working alongside communities to identify what aspects of heritage are meaningful, how they are to be represented, and in what ways tourism might contribute to their long-term viability. It also requires sensitivity to the pressures that global tourism can place on cultural systems, and a commitment to avoiding interventions that compromise authenticity or erode cultural agency. When applied thoughtfully, the UNESCO frameworks provide a foundation for ethical decision-making and long-term cultural stewardship in tourism development.

ACTIVITY 2: UNESCO IN PRACTICE – DESIGNING FOR CULTURAL SUSTAINABILITY

Purpose

In this activity you will apply the principles introduced in the UNESCO frameworks to design a THE experience that supports cultural sustainability. The activity invites you to consider how professional decisions can reflect the values of cultural diversity, safeguard living heritage, and promote inclusive cultural expression.

Instructions

Working individually or in pairs, complete the following task using a blank A3 worksheet or digital slide:

1 **Select a cultural practice or tradition** from a real or fictional community that could form the basis of a visitor experience. This might include a form of storytelling, ritual, seasonal food practice, craft technique or community celebration.

2. **Design a tourism experience** that centres on the chosen practice. Your design should include:
   - a brief outline of the visitor activity (what participants will do or learn)
   - the role of the host community in shaping and delivering the experience
   - key considerations to ensure the experience is respectful and sustainable
3. **Apply the UNESCO frameworks** using the following guiding questions:
   - How does your experience promote respect for cultural identity and diversity?
   - In what ways does it contribute to the safeguarding and transmission of intangible cultural heritage?
   - How does it support the creation, access or exchange of cultural expressions in a way that benefits the local community?
4. **Include a reflection** identifying one potential risk to cultural sustainability in your design. Suggest a practical strategy to address or minimize this risk.
5. **(Optional extension)**: Exchange your design with another individual or pair. Provide one comment on how their idea reflects UNESCO's principles and one suggestion for strengthening its cultural sustainability.

Challenges to cultural sustainability

While cultural heritage has significant value for communities and plays a vital role in shaping destination identity, it is increasingly subject to pressures that can undermine its sustainability. Within the tourism, hospitality and events (THE) industries, the promotion and consumption of cultural practices often occur within frameworks shaped by market demand, external interpretation and global visibility. These forces can lead to forms of engagement that distort, appropriate or marginalize the heritage they claim to celebrate.

Sustainable cultural tourism requires a critical understanding of the specific practices that can harm cultural sustainability, particularly when tourism is developed without sufficient input from the communities whose heritage is being represented.

Cultural appropriation

Cultural appropriation refers to the unauthorized or insensitive use of cultural elements by individuals or institutions outside the originating culture. These elements may include symbols, practices, artefacts or expressions that hold cultural significance. Gertner (2019) defines cultural appropriation as 'the unsuitable, unauthorized, or objectionable use of cultural elements in a context other than that of the culture by outsiders who might lack understanding and/or respect for the culture in question'. This becomes particularly problematic in tourism when cultural symbols are detached from their original context and repurposed for entertainment or profit without community involvement or consent.

Figure 4.2 illustrates the example of Hawaii as explored by Gertner, who looked at the tension between cultural representation and commercialization. While tourism contributes significantly to the state's economy, many Native Hawaiians express concern that their culture is misrepresented and commodified. Cultural performances, imagery and symbols are often used in ways that reinforce stereotypes and fail to reflect Indigenous perspectives. Surveys indicate that 60 per cent of Native Hawaiians believe tourism does not support the preservation of their culture or language. These perceptions reflect the long-term impacts of tourism development that fails to engage with host communities as equal partners.

Decontextualization and over-commodification

Bartolozzi (2023) describes decontextualization as the process by which cultural meaning is stripped from an item or activity, often reducing it to a commercial product.

Figure 4.2 The dilemma of the business of tourism and cultural appropriation

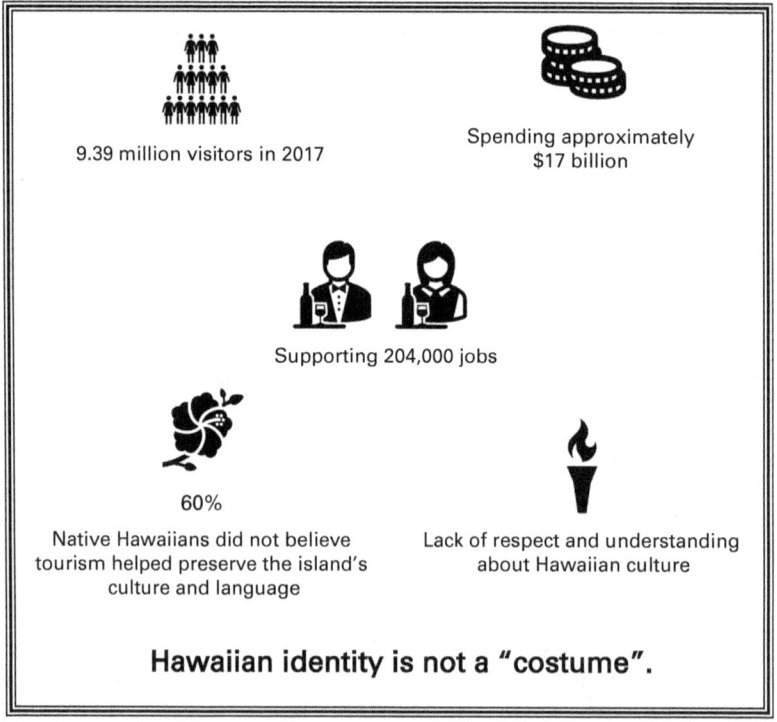

Data from Gertner, 2019

In many tourist settings, shopping becomes a primary means of engaging with culture. While purchasing a handmade item such as a textile or piece of pottery may support local economies, these transactions often occur without an understanding of the item's significance, production method or cultural origin. This shift from cultural engagement to consumer transaction can reinforce shallow interpretations of culture and place pressure on communities to produce stereotyped or simplified versions of their heritage to meet tourist expectations.

In response to these concerns, Butvin (2023) argues that tourism should create opportunities for visitors to engage with the people behind cultural expressions, not just the products. She emphasizes that cultural heritage is most meaningful when rooted in its original context and supported by community-led interpretation. This requires creating tourism experiences that involve explanation, participation and dialogue, rather than passive observation or consumption. Butvin warns that tourism can erode cultural meaning if industry practices prioritize market appeal over authenticity and community benefit.

Ethnocentrism and nationalism

These are further challenges that can distort cultural heritage in tourism contexts. Ethnocentrism involves viewing and interpreting other cultures through the lens of one's own cultural norms and values. In tourism, this can lead to presentations of heritage that reflect dominant perspectives or external standards of value rather than those of the community itself.

Hollinshead (2018) argues that ethnocentric thinking is a persistent issue in the THE industries, stating that 'ethnocentric outlooks are a prominent source of pain and error in the practice of tourism'. He notes that tourism professionals often lack the training or awareness to recognize how cultural representations may exclude alternative or subcultural voices. Nationalist frameworks can further complicate matters when heritage is used to promote a singular identity that marginalizes internal diversity or silences contested histories.

Authenticity and fake culture

One result of the dynamics already described is the emergence of staged or 'fake' culture. This occurs when cultural practices are replicated or performed outside their original context, often for the benefit of tourists, but without the involvement or leadership of the community from which they originate. Butvin (2023) and Jones (2023) describe how some tourism venues recreate traditional settings or rituals without community participation, presenting dramatized versions of heritage that may appear authentic but lack cultural legitimacy. In these cases, performers are often not members of the represented culture, and the performances are shaped more by tourist expectations than by community values. Such practices risk reducing complex cultural systems to entertainment, undermining the ability of communities to control how their heritage is shared and understood.

Butvin (2023) further explains that when majority groups or commercial operators take control of minority cultural representations, they may exclude the very communities who created and continue to practice those traditions. This not only displaces potential economic benefits but also disrupts the social functions and meanings attached to cultural practices. Over time, this can contribute to a loss of intergenerational transmission, weakening the sustainability of cultural heritage.

To support ethical and sustainable cultural tourism, Butvin (2023) and the Smithsonian Center for Folklife and Cultural Heritage propose a set of evaluative questions to guide the design of cultural experiences. These include:

- Is the experience community-based, and does it reflect community perspectives?
- Is it grounded in ethnographic or cultural research?
- Is cultural heritage central to the experience, and is it represented responsibly?
- Does the host demonstrate deep knowledge and the ability to discuss sensitive issues?
- Is the experience regularly updated to reflect evolving cultural practices and contexts?

These concerns also highlight the need for proactive planning in THE operations. Sustainable tourism requires that potential impacts on cultural heritage are considered from the earliest stages of development. Genuine and immersive experiences reflecting a community's ongoing cultural expressions can offer meaningful alternatives to superficial or exploitative forms of engagement. This means that in terms of sustainable development, THE organizations must recognize their responsibility to assess potential harms and design activities that prioritize cultural integrity.

Butvin (2023) emphasizes that cultural sustainability must begin during the idea and planning phase of tourism development. She identifies four key questions that can guide this process:

1 Are communities the primary beneficiaries of any tourism associated with their own intangible cultural heritage?
2 How does the activity promote their role as leaders in managing such tourism?
3 Can the viability, social functions and cultural meanings of intangible heritage be maintained without being diminished or threatened by tourism?
4 Is there an effective way to guide both industry interventions and the behaviours of those who participate as tourists?

These questions offer a practical framework for professionals seeking to embed sustainability principles into tourism design and implementation. They shift the focus from reactive mitigation to proactive safeguarding, ensuring that heritage is not only protected but supported through meaningful, community-led engagement.

Overtourism and its impact on cultural heritage

> **INDUSTRY VOICE: KAITLIN ARENS**
>
> Kaitlin Arens, resort experience ambassador at Edgewood Tahoe Resort, USA, says 'To tackle the challenges of over-tourism, it's vital to raise visitor awareness and foster understanding of local life. Meaningful engagement with the community lies at the heart of ethical and responsible tourism.'

Overtourism refers to a situation in which the number of tourists at a destination exceeds the capacity of the environment, infrastructure or community to cope with the negative effects. As defined by the World Tourism Organization (UNWTO), overtourism is 'the impact of tourism on a destination, or parts thereof, that excessively influences the perceived quality of life of citizens and/or the quality of visitor experiences in a negative way' (UNWTO, 2018). The concept encompasses both the experiences of residents, who may feel burdened by the tourist presence, and those of visitors, who may find that the high volume of tourists reduces the quality of their visit. For instance, international tourist arrivals have grown from 25 million in 1950 to over 1.3 billion in 2017, with forecasts estimating 1.8 billion annual cross-border tourists by 2030 (UN News, 2017).

Figure 4.3 illustrates the change in visitors by region from 2010 to 2020 and the projected number for 2030.

Although the term gained widespread recognition only after 2017, the underlying dynamics of overtourism are not new. The phenomenon is rooted in long-standing urban pressures, resource overuse, and tensions between hosts and guests (Koens, Postma and Mak, 2018). What distinguishes the current phase is the intensity of tourism flows and the role of digital and economic shifts such as sharing economy platforms and real estate speculation.

Distinguishing overtourism from mass tourism

Overtourism is often mistakenly equated with mass tourism. However, the two concepts are not synonymous. While mass tourism involves large volumes of visitors, overtourism refers to the negative effects perceived by residents and visitors when

Figure 4.3 International tourist arrivals by region (2010, 2020, 2030)

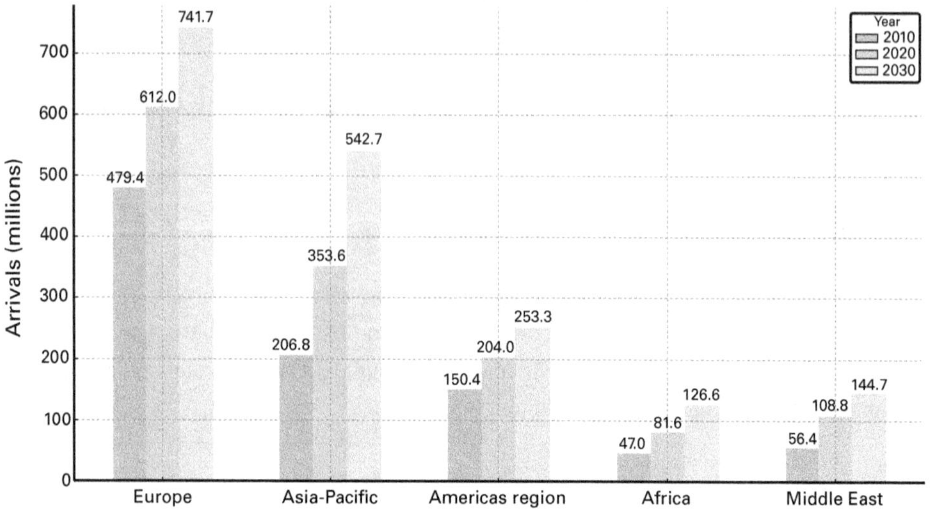

Region	2010	2020	2030
Europe	479.4	612.0	741.7
Asia-Pacific	206.8	353.6	542.7
Americas region	150.4	204.0	253.3
Africa	47.0	81.6	126.6
Middle East	56.4	108.8	144.7

Data from UNWTO, 2011

tourism activity, regardless of absolute numbers, disrupts everyday life, overburdens infrastructure or damages cultural and natural assets (Koens, Postma and Mak, 2018; UNWTO, 2018). This distinction explains why some destinations, such as London, can absorb high tourist numbers without being considered impacted by overtourism, while others like Venice or Dubrovnik face acute impacts despite receiving fewer visitors.

Importantly, overtourism tends to be highly localized. It is most commonly concentrated in specific areas, times or events such as historic quarters or peak seasons, yet can influence broader public sentiment and policy debates. In Barcelona, for example, intense tourism pressures in El Raval and the Gothic Quarter have contributed to city-wide concerns about housing, mobility and public space (Koens, Postma and Mak, 2018).

The causes of overtourism extend beyond tourism itself. Tourists compete with residents, commuters and day visitors for access to urban spaces and services. Meanwhile, broader societal changes such as digitalization, online shopping, social media, and real estate speculation have increased the pressure on and strengthened the view that the quality of life for residents is worsening. The issue of overtourism is social and environmental and requires inclusive, context-sensitive responses (Koens, Postma and Mak, 2018).

Overtourism can significantly affect both tangible and intangible cultural heritage. In Dubrovnik, Croatia, the medieval city walls have deteriorated under the weight of excessive tourist footfall, prompting restrictions on cruise ship arrivals (able2travel, 2024). In Venice, large numbers of tourists, particularly during peak seasons, have strained the capacity of transport, sanitation and historic infrastructure. In response, a day-tripper entry fee was introduced in 2024 (Calder, 2025).

Cultural environments are also reshaped as traditional services give way to visitor-oriented commerce. In Barcelona's historic core, rising rents and demand for tourist accommodation have displaced long-standing residents and small businesses (Milano, 2017). While these effects may be concentrated in particular areas, their influence extends across the urban and cultural landscape.

Informal sector workers such as artisans, street vendors and performers can be impacted by the change. These locals are often excluded from decision-making and lack access to the protections afforded to formal businesses, even as their livelihoods are tightly linked to heritage tourism.

Social tensions and resident displacement

Overtourism exacerbates social tensions by disrupting the daily lives of residents in both urban and rural destinations. These tensions manifest through overcrowded public spaces, increased cost of living, and the erosion of local cultural identity. The influx of tourists often leads to a sense of loss among residents, who feel their communities are being transformed into commodities for external consumption (Pechlaner, Innerhofer and Erschbamer, 2022).

Digital platforms, notably Airbnb, have intensified these issues by facilitating the rapid conversion of residential properties into tourist accommodations. This shift not only reduces the availability of housing for locals but also alters neighbourhood dynamics, leading to feelings of alienation among long-term residents (Pham, Andereck and Vogt, 2024).

Governance challenges further compound these social tensions. In many cases, local authorities struggle to implement effective policies to manage tourist flows and mitigate their impacts. The lack of coordinated strategies often results in reactive measures that fail to address the root causes of overtourism, leaving communities feeling disempowered and voiceless in decisions affecting their environment (Pechlaner, Innerhofer and Erschbamer, 2022).

Community resilience emerges as a critical factor in responding to overtourism. Pham, Andereck, and Vogt (2024) highlight the importance of involving local stakeholders in tourism planning and decision-making processes. By developing inclusive governance structures, communities can develop adaptive strategies that balance the needs of residents with the demands of tourism, thereby enhancing social cohesion and preserving cultural integrity.

Environmental and infrastructure pressures

Overtourism contributes significantly to the degradation of both environmental quality and urban infrastructure. As tourist numbers grow, pressure increases on transport systems, water supply, waste management and energy networks – often beyond what these systems were originally designed to handle. In many destinations, especially those with heritage value or ecologically sensitive landscapes, such demand accelerates the wear on public utilities, damages fragile ecosystems and depletes local resources (Pechlaner, Innerhofer and Erschbamer, 2022).

Environmental impacts are not confined to urban areas. In rural and island destinations, overtourism can lead to the overuse of coastal zones, degradation of forests and waterways and unsustainable waste generation. These pressures are amplified during peak seasons, creating a cyclical pattern of environmental stress. According to Pham, Andereck and Vogt (2024), unmanaged tourism can also contribute to biodiversity loss and habitat fragmentation, particularly in areas where tourism infrastructure encroaches upon protected or agricultural land.

Infrastructure overload is often most visible in transport and sanitation systems. In cities such as Venice and Dubrovnik, daily surges of cruise ship passengers strain transit networks, crowd narrow streets and overwhelm public sanitation. In island contexts, such as Bali or Santorini, water shortages and waste disposal challenges have emerged as critical concerns due to tourism-driven overconsumption and limited local capacity for environmental management (Pechlaner, Innerhofer, and Erschbamer, 2022).

Technological responses – such as smart traffic systems, digital waste monitoring or tourism apps – can support destination management, but they are not solutions. As Pham, Andereck and Vogt (2024) argue, environmental degradation stems from the structural imbalance between short-term tourism growth goals and long-term sustainability planning. Without coordinated regulation, the physical footprint of tourism will continue to expand beyond ecological and infrastructural limits.

To mitigate these pressures, research emphasizes the importance of integrated destination planning. Strategies include setting ecological carrying capacity thresholds, investing in green infrastructure, and aligning tourism development with climate adaptation and environmental protection goals. Such measures require political will and cross-sector collaboration to ensure that both resident wellbeing and environmental integrity are prioritized.

Strategies and responses to overtourism

Addressing overtourism effectively requires the integration of both strategic governance and operational management. In 2017, McKinsey & Company, in collaboration with the World Travel & Tourism Council (WTTC), proposed four high-level governance priorities: evidence-based policymaking, long-term planning, cross-sector collaboration and financial transparency.

Building on this framework, the World Tourism Organization (UNWTO, 2018) introduced a set of 11 actionable strategies aimed at helping destinations manage visitor pressure and reduce negative impacts. These measures focus on practical aspects such as destination management, community engagement, visitor behaviour, infrastructure planning and the use of real-time data.

While the UNWTO strategies are tactical in nature, their effectiveness is enhanced when implemented within the broader governance structure outlined by McKinsey & Company. Together, these approaches provide a flexible and scalable framework for managing overtourism at both local and systemic levels.

ACTIVITY 3: MAPPING UNWTO STRATEGIES TO MCKINSEY GOVERNANCE PRIORITIES

Purpose

Understand how practical tourism management strategies (UNWTO) align with broader governance priorities (McKinsey & Company) in addressing overtourism.

Instructions

In the table below you have been given the UNWTO 11 strategies for addressing overtourism and the rationale for each strategy. You have also been given McKinsey & Company's broader governance priorities in addressing overtourism.

Match each UNWTO strategy with the most relevant McKinsey governance priority. Some strategies may align with more than one priority; use the rationale to help you to decide. To get you started, one has been done for you.

UNWTO strategy	Aligned McKinsey & Company governance priority	Rationale
1. Disperse visitors across attractions and geographic areas		Spatial dispersion requires long-term destination planning and land-use management.
2. Promote off-peak travel		Temporal dispersion aligns with demand management and economic sustainability goals
3. Develop niche and alternative tourism routes		Encourages market diversification and relieves pressure on iconic sites.
4. Enforce regulations on tourism	**Build and update a robust evidence base**	**Effective enforcement depends on clear data, monitoring and legal frameworks.**

(continued)

(Continued)

UNWTO strategy	Aligned McKinsey & Company governance priority	Rationale
5. Target specific market segments		Segmentation helps shape visitor profiles aligned with destination goals.
6. Ensure benefits to local residents		Emphasizes equitable tourism and community-based economic inclusion.
7. Design mutually beneficial experiences for tourists and residents		Requires collaboration between public, private and community actors.
8. Expand infrastructure capacity responsibly		Infrastructure upgrades often demand external or diversified financing models.
9. Engage local communities in policymaking		Reflects participatory governance and inclusive decision-making.
10. Educate tourists about their impacts		Promotes behavioural change through collaborative communication and stakeholder alignment.
11. Use real-time data to monitor tourism trends ad guide policy		Real-time analytics are fundamental to evidence-informed governance and rapid response.

McKinsey & Company's four governance priorities:
1. Build up and update a robust evidence base
2. Plan for sustainable tourism growth
3. Involve all sectors of society
4. Secure new and transparent funding sources

The alignment of operational strategies and governance priorities provides a strong foundation for managing overtourism. However, the real challenge lies in applying these tools in specific destinations with unique environmental, social and political contexts. A good example is South Lake Tahoe, Sierra Nevada, where tourism growth has resulted in traffic congestion, housing shortages and resident dissatisfaction. In response, local authorities have implemented a mix of UNWTO-style strategies, including rental caps, infrastructure upgrades, and visitor education campaigns, supported by governance actions such as data monitoring and inter-agency collaboration. This case illustrates how theory and strategy converge in practice, offering important lessons for sustainable destination management.

The integration of UNWTO's operational strategies with McKinsey's governance priorities provides a structured approach to addressing overtourism. However, the effectiveness of these frameworks depends on their adaptability to specific destination contexts. The following real-world example of South Lake Tahoe examines how one destination has navigated the complexities of overtourism, offering insight into the challenges and opportunities of applying these strategies in practice.

REAL-WORLD EXAMPLE South Lake Tahoe, Sierra Nevada, USA

Background

South Lake Tahoe (SLT), a popular tourist destination nestled in the Sierra Nevada mountains, faces significant challenges due to over-tourism and its delicate alpine environment. Despite a population of 21,330 people (US Census Bureau, 2020), Lake Tahoe attracts between 15-18 million annual visitors (Boger and DeSilva, 2023). This influx of tourists has led to a variety of issues, including increased pollution, traffic congestion and strained local infrastructure. One of the most pressing concerns is the risk of catastrophic fires and visitor entrapment. Pollution, climate change and increased human activity have contributed to higher hazardous fuel ignition rates, making the region more susceptible to fires, and posing a significant threat to the water quality of its fragile ecosystem (Lake Tahoe Basin Management Unit, 2009). This demonstrates an excellent example of the need for enhanced sustainability in tourist destinations like SLT.

The pressure of popularity

Overtourism and its environmental, sociocultural and economic implications have led to increased scrutiny of the event industry. As the industry seeks to minimize negative impacts and maximize positive contributions, sustainability has become the event organization's utmost concern (Bladen et al, 2023). While events play a crucial role in SLT's economy, they can also contribute to environmental degradation and social problems. The indirect impact of events on local communities, both positive and negative, highlights the importance of event sustainability especially in tourist destinations (Smith and Mair, 2021). Thus, the threat of mass tourism can lead to higher living costs, inflation, crime and other issues (McCool and Bosak, 2021). To mitigate these risks, tourists, businesses and the community must understand how responsible tourism can promote environmental and social preservation. By fostering a sense of responsibility for tourism initiatives, SLT can ensure that tourism remains a force for positive change.

When the local community of SLT was surveyed, they revealed that the areas of their life that had been most affected by tourism were pollution (45 per cent), traffic (78 per cent) and increased population (34 per cent). These issues are common indicators of overtourism in destinations.

The role and responsibility of events

Local businesses also recognized the importance of sustainability, particularly in waste reduction and vendor practices. They acknowledged the potential negative impacts of events on the community and emphasized the need for education and awareness about Lake Tahoe's fragile and delicate ecosystem. By encouraging tourists to adopt sustainable practices and maintain clean facilities, businesses aim to contribute to a healthier community and protect the lake's pristine waters. Education is very important for success in sustainable development, not only for events but also for the destination. Thus, enhancing visitor awareness and understanding of community life are key factors to begin addressing the challenges of over-tourism in SLT (Lovato, 2023). This prioritization allows companies to demonstrate their commitment to sustainable development, benefiting society, the environment and the economy while minimizing the negative effects events have on the local community.

Community perception and business engagement

A strong relationship between businesses and the community is vital for successful sustainable initiatives. However, the survey found that 79 per cent of residents perceived event companies as contributing little to nothing to the local community, likely due to insufficient communication about their engagement efforts. Events offer opportunities for resource exchange, dialogue, transparency and trust-building, which are crucial for building a positive relationship with stakeholders and the local community (Orefice and Nyarko, 2021). Engaging with the community is vital for building support and avoiding negative consequences. Without such engagement, residents may distrust authorities and sustainability efforts, impacting their quality of life.

Barriers to sustainable implementation

Implementing sustainability in the events industry comes with its challenges. In SLT, their main issue pertains to convincing clients to adopt sustainable practices due to costs and potential conflicts with their vision for the event. Effective communication about the benefits of sustainable events is crucial for persuading clients. Additionally, local regulations, such as restrictions on certain power products for exhaust emissions and stricter permits, can pose challenges for event managers due to the increased costs of maintaining a clean environment and ecosystem. Businesses must navigate these hurdles while emphasizing and communicating the advantages of sustainability and its positive impact on the community contributing to an improved quality of life.

Legacy and opportunity

While over-tourism remains a significant challenge, SLT businesses continue to make efforts to engage with the local community. These include group events, youth outreach programmes, charity work, and hiring local vendors rather than corporate ones. Engaging with the community is essential for ethical and responsible tourism, enhancing a company's corporate social responsibility (CSR) image (Carroll, 2021). A large majority of residents are unaware of these efforts, and this has led to a perception that event tourism only has negative impacts. To address this, businesses need to improve their advertising, CSR practices and communication to ensure the public visibility of their sustainability initiatives, preserving the future legacy of SLT events.

The current legacy of events in SLT is a subject of varying perspectives. While many tourists consider the area a top destination for weddings and events, locals have a different view of the impacts of events on their community. Sustainability has become increasingly important for tourism economies, presenting numerous opportunities and challenges for the events industry. To meet rising expectations for environmentally responsible events, organizers must focus on the legacy and impact of their events, ensuring they meet sustainability goals while preserving the environment. By implementing a sustainable growth strategy, SLT can amplify its positive impacts of events, enhancing community engagement, business reputation and environmental consciousness. This ultimately benefits the local community and establishes a positive legacy for event businesses.

Smith and Mair (2021) suggest that events can positively contribute to sustainable development by educating and inspiring people to adopt sustainable lifestyles. However, SLT's popularity creates a complex situation. While tourism fuels the economy, it also strains the environment and local resources. The key lies in sustainable practices. Despite challenges, event businesses show commendable commitment to sustainability. While there are notable negative impacts, events in SLT are generally perceived positively. Thus, by working together, businesses, tourists and the community can find ways to minimize negative impacts and maximize positive ones. This includes educating visitors, fostering responsible event management and strengthening communication between businesses and residents. By prioritizing sustainability, SLT can ensure tourism remains a force for good, protecting its natural beauty and fostering a thriving community.

ACTIVITY 4: CONFRONTING THE CHALLENGES OF OVERTOURISM

Purpose

This activity asks you to critically reflect on the sustainability challenges presented in the South Lake Tahoe real-world example. You will apply management principles to design more responsible tourism and event practices.

> *Instructions*
>
> 1. **Identify and summarize** one key environmental, social, and economic challenge faced by South Lake Tahoe due to overtourism.
> 2. **Choose ONE of the challenges** and propose an idea that a local THE business could use to respond to the challenge in a sustainable way.
> 3. Consider **who would need to be involved** for the strategy to succeed, for example stakeholders, operational teams or partnerships.
> 4. Reflect on **how your strategy balances the needs** of the business, the community and the environment.
> 5. **Prepare a short summary** of key points.
>
> **Peer discussion (optional)**
>
> With your classmate, share your ideas and discuss:
>
> - Which types of challenges came up most often? Why you think that occurred?
> - Which ideas are most realistic or creative?
> - How could THE work more closely with the local community to address the challenges?

Corporate social responsibility and stakeholder theory

The discussion in this chapter so far has looked at the importance of tangible and intangible cultural heritage as a sustainability issue in which THE has a role to play. It is within the frameworks of corporate social responsibility (CSR) and stakeholder theory that THE's business role can be understood.

CSR and stakeholder theory are interconnected, offering complementary frameworks to address the complex challenges of sustainability and business growth. THE businesses, ranging from multinational hotel chains to small eco-tourism ventures, operate within a network of interdependent stakeholders, including tourists, local communities, employees, suppliers, government and the environment. Aligning CSR and stakeholder theory in THE provides a strategy for addressing diverse responsibilities while ensuring long-term value creation (Freeman and Dmytriyev, 2017).

CSR has been defined as 'the idea that businesses should operate according to principles and policies that support sustainability by making a positive impact on society and the environment' (McGrath and Jonker, 2023). It is important because it requires businesses to consider corporate citizenship and the impact of their business on society.

Stakeholder theory was introduced by Freeman and McVea (2001) who argued that businesses needed to shift toward values-based management that recognizes the economic, social, political and ethical considerations of operations. As a concept it was developed at the same time as UNESCO was focussing upon cultural diversity, cultural expression and cultural heritage as issues of sustainability. Freeman and McVea outlined seven key features of the stakeholder approach:

1. It is a single strategic framework for navigating changing external environments without needing new models.
2. It utilizes a strategic *management* process to replace a strategic *planning* process so that strategy formulates rather than predicts the future.
3. It implements objective alignment with those impacted by or capable of influencing operations to achieve organizational goals.
4. It adopts a values-based management approach. A values-based management approach aligns an organization's actions, decisions and strategies with its core values. It should emphasize the nurturing of long-term cooperation among diverse stakeholders to guide its behaviour, decision-making and interactions.
5. It takes a holistic approach that combines economic, social, political and ethical considerations for values-based management.
6. It acknowledges stakeholder specificity by holding a detailed understanding of stakeholders to create tailored, effective solutions.
7. It is sustainable by balancing multiple stakeholder interests to ensure continued business success.

Freeman and Dmytriyev (2017) aligned CSR with stakeholder theory, stating that there were common elements of purpose, value creation and stakeholder interdependence. These elements will be explored as they apply to THE.

Purpose

According to Freeman and Dmytriyev (2017), an organization's purpose should be ethically rooted. It should guide the organization to create value for all stakeholders while contributing positively to society at large. Its vision, mission and strategy should shape its CSR. In their view, a purpose-centred approach integrates societal, environmental and economic objectives, countering false oppositions such as 'economic vs social' or 'profits vs ethics'.

In THE, purpose is integral to aligning business operations with the broader needs of the host community and the environment. THE businesses often have unique opportunities to enhance cultural understanding and promote sustainable development while driving economic growth. For example, a THE company could partner with local communities to create authentic cultural experiences, and support conservation efforts. Such a purpose ensures that tourist activities generate value not only for tourists and the business itself, but also for the local economy, its people and cultural heritage.

Value creation

For Freeman and Dmytriyev (2017), value creation is central to stakeholder theory and CSR because it emphasizes the need for business to positively impact all their stakeholders. They point out that value creation is not limited to financial gains but extends to creating social, environmental and economic value. By recognizing the interdependence of stakeholders, value creation encourages organizations to make decisions that simultaneously benefit customers, employees, suppliers and communities. It rejects the view of 'trade-offs' among stakeholders, instead advocating for integrated strategies where the success of one group contributes to the wellbeing of others.

Value creation in THE can significantly contribute to the preservation of cultural heritage by aligning business strategies with the principles of stakeholder engagement. This approach ensures that THE benefits both the social and cultural fabric of host destinations. For example, when considering cultural heritage preservation, revenue from tourist activities, such as entrance fees to heritage sites or local representations of traditions, can be directed towards maintenance and restoration of historical landmarks and local treasures.

From a community empowerment and environmental sustainability perspective, THE organizations could ensure that their employees are versed in community-owned and ethically managed enterprises operating within the region. For example, a hotel in a region that attracts visitors for whale-watching could ensure that it directs its guests to operators who adhere to the guidelines of the International Whale Commission and Department of Conservation, and are certified by the World Cetacean Alliance.

Stakeholder interdependence

Freeman and Dmytriyev (2017) are critical of the prevailing idea in business that decisions often involve trade-offs due to limited resources. For instance, helping communities may seem to lower shareholder returns, offering better supplier terms might imply higher costs for customers, or providing premium employee compensation could reduce value for other stakeholders. However, they argue that these are false perspectives that oversimplify the reality of stakeholder relationships and miss an essential point: stakeholders are interdependent. Actions that benefit one group often create ripple effects of value across others.

In THE, stakeholder interdependence is particularly evident, given its reliance upon a complex network of local suppliers, transportation providers, community engagement, tour operators, event organizers, government, and related businesses such as cleaning and laundry services, visitors and the environment.

Stakeholder interdependence is evident throughout all types of THE operations, even in less conventional or unexpected settings. By exploring atypical examples, the interplay between stakeholders often takes unique forms.

CSR extends beyond legal requirements to include the implementation of ethical, sustainable and responsible business practices for corporate citizenship and broader societal impact in decision-making. It is through ethical operations that THE can address its social and cultural obligations (McGrath and Jonker, 2023).

REAL-WORLD EXAMPLE The IceHotel, Jukkasjärvi, Sweden

The IceHotel in Sweden offers one perspective on CSR, values-based management and stakeholder theory in practice. As you read about the IceHotel, consider the hotel's purpose and value creation for its stakeholders.

Background

The IceHotel in Sweden is constructed every winter from the ice formed by the Torne River. It is situated on the Torne River in the historic marketplace of Jukkasjärvi in Sweden. It offers the experience of staying in accommodation made entirely of ice and snow. Its purpose is to offer a unique, sustainable and culturally immersive experience while preserving the natural and cultural heritage of its surroundings.

Local government interdependence

The IceHotel collaborates with local authorities to promote Arctic tourism. This contributes to the region's income while adhering to strict environmental regulations. The IceHotel is certified as a Sustainable Arctic Destination which requires the operations to be sustainable throughout areas of sustainable management, socio-economic impact, cultural impact and environmental impact (Kiruna Lappland Economic Association, 2022).

Environmental interdependence

Every year in the winter months the Torne River turns to ice. The ice is harvested and stored and eventually used to build the IceHotel. In the summer months the hotel melts and returns

Figure 4.4 The IceHotel, Jukkasjärvi, Sweden

Photographer: Stephan Herz, 2006, www.sherz.net

to the river. The amount of wastage and emissions is low as the ice is not transported from its source, and any unused ice is returned to the river. It is run on 100 per cent renewable energy (IceHotel, .n.d.).

Local community interdependence

The hotel prioritizes sourcing its food and beverage ingredients from local herders, fisheries and farmers, and forages from local nature. The favoured produce is organic and responsibly procured. Visitor excursions and experiences have an educational focus, often including engagement with the Sámi culture (IceHotel, .n.d.).

Other stakeholder interdependence

The IceHotel management collaborate with international and local artists to annually create the new hotel. Artists gain global exposure and royalties, while guests experience unique and annually changing art installations (IceHotel, .n.d.).

ACTIVITY 5: UNDERSTANDING STAKEHOLDER INTERDEPENDENCE AT THE ICEHOTEL

Purpose

In this activity, you will explore how the IceHotel demonstrates stakeholder interdependence and values-based tourism. You will identify the hotel's key stakeholders and reflect on how its operations align with sustainability principles.

Instructions

1 **Read the IceHotel real-world example**
2 **Use the table below to identify**:
   o one stakeholder group (e.g. local community, environment, local government, visitors, artists)
   o one way in which the IceHotel creates value for this group
   o one reason why this value matters for cultural or environmental sustainability
3 **Complete the table** with at least three stakeholder groups. An example has been done for you.

Stakeholder group	How the IceHotel creates value	Why this matters for sustainability
Local community	Sources food from Sámi herders and local producers	Supports local livelihoods and cultural identity

Designing ethical cultural experiences: Aligning practice with principle

It is evident from the real-world examples of South Lake Tahoe in the USA and the IceHotel in Sweden that a stakeholder-driven model of cultural sustainability is key. They recognize that part of their CSR obligation is to acknowledge that the design of experiences has to be sustainable, inclusive and community-led.

Earlier in this chapter Butvin's (2023) two sets of guiding questions for professionals working in THE were outlined: one set of questions was concerned with planning, and the other with evaluation. They are reiterated here in full because, in the context of design, they perform a distinct function: they help ensure that sustainability and cultural responsibility are embedded from the outset and maintained throughout implementation.

The first set of questions (Butvin, 2023) guides the initial design of cultural tourism experiences. Their purpose is to shape decision-making before a product is finalized:

1 Are communities the primary beneficiaries of any tourism associated with their own intangible cultural heritage?

2 How does the activity promote their role as leaders in managing such tourism?

3 Can the viability, social functions and cultural meanings of intangible cultural heritage be maintained without being diminished or threatened by tourism?

4 Is there an effective way to guide both industry interventions and the behaviours of those who participate as tourists?

These questions centre the values and interests of cultural communities. They require professionals to consider who holds power in the design process, how heritage is interpreted, and what safeguards are in place to ensure that practices remain meaningful and viable. Applied at the planning stage, they help prevent harm by making equity, context and accountability foundational design considerations.

The second set of questions provides criteria for evaluating cultural experiences, whether during design development or in post-implementation review. While already discussed in the context of ethical assessment, they are equally valuable during design for refining and improving content and delivery:

1 Is the experience community-based, and does it reflect community perspectives?

2 Is it grounded in ethnographic or cultural research?

3 Is cultural heritage central to the experience, and is it represented responsibly?

4 Does the host demonstrate deep knowledge and the ability to discuss sensitive issues?

5 Is the experience regularly updated to reflect evolving cultural practices and contexts?

These questions support reflective design by encouraging alignment with cultural realities, practitioner knowledge and heritage integrity. They help determine whether the experience communicates cultural meaning appropriately, whether the people delivering it are adequately informed, and whether the structure of the experience adapts to change.

Together, these two sets of questions form a practical framework for ethical cultural experience design. The planning-stage questions encourage shared authority and community-defined purpose, while the evaluative questions offer a mechanism for ensuring that these intentions are maintained over time. When used in parallel, they enable professionals to design cultural experiences that are locally rooted, ethically structured and sustainable in practice.

This approach draws directly from the values embedded in UNESCO's key frameworks: the Universal Declaration on Cultural Diversity (2001), the Convention for the Safeguarding of the Intangible Cultural Heritage (2003), and the Convention on the Protection and Promotion of the Diversity of Cultural Expressions (2005). Each framework affirms the right of communities to protect, express and renew their cultural life. The role of the THE professional is to respect and support these rights through meaningful collaboration, informed decision-making and culturally responsive delivery.

Design processes must also recognize that heritage is not fixed. It changes with communities and through time. Practices that are shared with visitors may evolve in form, language or significance. Ethical cultural experience design supports this evolution by providing flexible and participatory structures that allow for cultural expression to remain relevant and alive. This includes not only decisions about what is shared, but also how, by whom, and under what conditions.

By integrating both planning and evaluation into the design process, THE professionals can move beyond compliance and toward values-based practice. This requires attention to power, accountability, and the lived experience of heritage within communities. Through this dual lens, design becomes a space for reinforcing cultural integrity and supporting community agency in the context of THE.

Aligning cultural sustainability with the Sustainable Development Goals (SDGs)

Cultural sustainability plays a key role in achieving various SDGs, yet there is no one single SDG that addresses culture. Culture plays a role in connecting and supporting all 17 SDGs, their targets and indicators (Hosagrahar, 2023). The goals promote the preservation of cultural heritage, the inclusion of diverse cultural practices, and the empowerment of local communities. Below are the SDGs that relate to cultural sustainability and how they impact THE practices.

Target 4.7 and Indicator 4.7.1

Figure 4.5   SDG 4

This target and indicator highlight the importance of integrating sustainable development principles, including understanding of cultural diversity and heritage in all types of education including lifelong learning and professional development. It measures how well global citizenship education and education for sustainable development are being met.

For THE professionals, this means acquiring the competencies, knowledge and understanding to devise and implement responsible, sustainable employment practices.

Target 8.9 and Indicator 8.9.1

Figure 4.6 SDG 8

This target calls for the promotion of sustainable tourism that creates jobs and supports local culture. It encourages tourism to contribute both the economic growth and cultural preservation. It measures the economic impact of tourism, including its contribution to cultural industries.

For those working in the THE sector, this target demonstrates how host communities can be supported economically and culturally by sustainability promoting local tangible and intangible heritage.

Target 10.2 and Indicator 10.2.1

Figure 4.7 SDG 10

Target 10 focuses upon empowering and promoting social, economic, and political inclusion for all, irrespective of age, gender, ethnicity, race, disability, religion or economic status. It measures the proportion of people living below 50 per cent of the average income.

From a THE perspective, SDG 10 requires professionals ensure inclusion in its practices by providing equal opportunities and respecting the voice and diversity of cultural identity within the host community.

Target 11.4 and Indicator 11.4.1

Figure 4.8 SDG 11

Target 11 focus on protecting and safeguarding cultural and natural heritage, particularly in urban environments. It stresses the need for urban planning that incorporates cultural preservation alongside development. It measures the resources spent on protecting and conserving that heritage.

THE professionals need to understand that this target is key when working in urban tourism. It is incumbent upon the THE business to ensure that its growth within cities does not come at the expense of cultural heritage.

Target 12.b and Indicator 12.b.1

Figure 4.9 SDG 12

Target 12.b is an outcome target focussing on creating and implementing strategies for sustainable tourism. It encourages responsible tourism that promotes local culture, reduces environmental impact, and supports sustainable production practices. It does this by tracking the number of countries with sustainable tourism strategies and action plans.

In any THE career, professionals will increasingly play a role in advocating for responsible consumption and production.

Target 16.7 and Indicator 16.7.1

Figure 4.10 SDG 16

This target is concerned with inclusive decision-making. It aims to ensure all communities, especially those with unique cultural traditions have a voice in shaping development policies. It measures how well populations feel included in decisions that affect their communities.

For aspiring THE professionals entering the industry, this target exemplifies the importance of engaging local communities in operational planning and decision-making to ensure cultural heritage and sensitivities are respected and preserved.

Target 17.16 and Indicator 17.16.1

Figure 4.11 SDG 17

The UN have stated that without SDG 17, it is very difficult to achieve the SDGs. This is because this target promotes partnerships and collaboration between stakeholders, including government, businesses and communities, to achieve sustainable development. It encourages the sharing of knowledge, resources and expertise. The implications for THE is that professionals must engage in transparent communication to ensure that business initiatives are co-designed with communities, allowing them to shape the development of cultural tourism that reflect the host's values and aspirations. In this regard, THE is contributing to the protection of cultural heritage while ensuring the benefits are equitably shared.

By utilizing the SDGs, THE can play a crucial role in the design of culturally sustainable experiences through community engagement and professionalism. SDG 17 accentuates the need for strong partnerships across sectors, urging the THE industry to work collaboratively with governments, local communities and other industry representatives. Collectively, the SDGs stress the importance of a THE model that respects cultural heritage, supports local economies and establishes partnerships.

Implications for the future of cultural tourism

This chapter has approached the sustainability of cultural tourism through a problem–solution logic, identifying overtourism as a key challenge that compromises the integrity of both tangible and intangible heritage. It has explored how tourism, when poorly managed or inadequately contextualized, can erode cultural meaning, displace communities and undermine the viability of cultural practices. In response, the integration of UNESCO frameworks, corporate social responsibility (CSR), and stakeholder theory has been proposed as a means of embedding ethical responsibility, participatory governance and long-term value creation into tourism practice.

Looking ahead, the future of cultural tourism will depend on the extent to which these frameworks are translated into operational norms. There is an increasing need to recognize heritage not as a static product but as a dynamic system of knowledge, identity and practice. This recognition necessitates a shift from extractive tourism models toward those that prioritize local agency, cultural continuity and shared decision-making. THE professionals will need to support environments in which communities determine how their heritage is represented, interpreted and transmitted.

CSR and stakeholder-based approaches offer strategic foundations for this transition. When aligned with sustainability objectives, these models enable THE enterprises to balance economic goals with the protection of cultural and social assets. Future cultural tourism initiatives will likely require strengthened stakeholder networks, transparent benefit-sharing mechanisms, and governance arrangements that ensure communities are not only participants but decision-makers.

In parallel, the growing urgency of addressing overtourism suggests that adaptive management strategies such as spatial and temporal dispersion, infrastructure reform and regulatory oversight will become more central to policy and planning. However, such measures must not be seen in isolation from cultural considerations. Sustainable destination management depends not only on regulating visitor numbers but on cultivating conditions in which cultural heritage remains viable, meaningful and accessible to those who produce and sustain it.

The implications for cultural tourism are therefore structural and systemic. A transition toward genuinely sustainable practice requires reconfiguring how tourism relates to place, people and heritage. It demands approaches that are rooted in equity, designed with community input, and responsive to local context. These principles are not optional; they are prerequisites for ensuring that cultural tourism continues to support rather than threaten the diversity and vitality of the world's cultural landscapes.

Conclusion

Cultural sustainability is an essential component of responsible THE practices. It requires an understanding of the complex relationship between the role of UNESCO's Declaration and Conventions on cultural heritage and diversity, the UN Sustainable Development Goals and business frameworks of corporate social responsibility and stakeholder theory. Culture plays a cross-cutting role across the SDGs, emphasizing the preservation of heritage, the inclusion of diverse practices and the empowerment of local communities. The SDG targets provide actionable pathways for THE professionals to engage in sustainable practices, from promoting cultural education and responsible tourism to fostering inclusive decision-making and collaborative partnerships.

The integration of CSR and stakeholder theory further supports the role of THE in creating value that benefits not just business, but all stakeholders including local communities and the environment itself. This values-based management approach highlights the interconnectedness of stakeholders and the importance of aligning THE operations with ethical and sustainable goals. By prioritizing stakeholder interdependence and promoting initiatives that respect and celebrate cultural heritage, THE businesses can contribute to both economic growth and cultural preservation.

The chapter has also explored critical challenges such as cultural appropriation, ethnocentrism and over-commodification, all of which pose a threat to the sustainability of cultural heritage. THE professionals must address these challenges by working towards authentic, community-driven practices that empower host communities as stewards of their heritage. Strategies such as adopting inclusive planning and culturally contextual professional development can mitigate harmful practices and promote genuine cultural exchange.

The role of THE extends beyond economic imperatives to include ethical, transparent and culturally sensitive practices. By actively engaging communities as partners in shaping local tourism, THE can promote cultural sustainability while contributing to global development. Through collaboration, education and inclusive decision-making, THE will be challenging prevailing ideas of decisions requiring trade-offs to ensure long-term industry success.

KEY TAKEAWAYS

- Culture as a dimension of sustainability:
    - Cultural heritage is recognized as one of the four pillars of sustainability, alongside environmental, social and economic factors. Its preservation is essential for maintaining identity, continuity and human dignity within communities.

- Ethical and community-based design of tourism experiences:
  - The design of tourism experiences should prioritize the involvement of host communities, ensuring that cultural practices are represented accurately and respectfully. Butvin's framework offers guiding questions to support ethical planning and evaluation.
- Negative impacts of tourism on cultural heritage:
  - Tourism activities can undermine cultural sustainability when they involve cultural appropriation, over-commercialization or the marginalization of local communities. These risks are heightened when community perspectives are excluded from the development process.
- UNESCO frameworks for cultural sustainability:
  - The 2001 Universal Declaration on Cultural Diversity, the 2003 Convention for the Safeguarding of the Intangible Cultural Heritage, and the 2005 Convention on the Protection and Promotion of the Diversity of Cultural Expressions provide international frameworks for protecting cultural heritage. These frameworks inform policy and professional practice within the THE industries.
- Corporate social responsibility and stakeholder theory:
  - These frameworks promote a values-based approach to management that considers the interests of all stakeholders. In the context of THE, this includes aligning business practices with social, cultural and environmental responsibilities.
- Cultural sustainability and the Sustainable Development Goals (SDGs):
  - Cultural sustainability is integrated across several SDGs, including those related to education, inclusion, heritage protection and partnerships. THE professionals are expected to contribute to these goals through practices that support local culture and community engagement.

REFLECTIVE QUESTIONS

1. How can tourism professionals ensure that cultural heritage is presented in a way that respects the values and perspectives of the host community?
2. In what ways might tourism activities unintentionally contribute to the loss or distortion of intangible cultural heritage?
3. What are the responsibilities of THE professionals when applying international frameworks such as the UNESCO conventions in local tourism contexts?

4 How can corporate social responsibility and stakeholder theory be used to balance the economic goals of tourism businesses with the cultural and social needs of local communities?

5 Reflect on a cultural tourism experience you have encountered or studied. To what extent did it promote cultural sustainability, and what could have been improved?

Further resources

Falter, M (2024) The role of rural tourism lifestyle entrepreneurs in rethinking current tourism development, *Journal of Tourism Futures*, https://doi.org/10.1108/JTF-09-2023-0205 (archived at https://perma.cc/6WF8-A5C8)

References

Able2travel (2024) 5 stunning locations where cruise ships are banned, 23 August, www.able2travel.com/news/5-stunning-locations-where-cruise-ships-are-banned/ (archived at https://perma.cc/HF2S-WMYD)

Bartolozzi, I (2023) *Heritage and Commerce: Rethinking cultural value in tourism*, Firenze University Press, Florence

Bladen, C, Kennell, J, Abson, E and Wilde, N (2023) *Events Management: An introduction*, 3rd edn, Routledge, London

Boger and DeSilva (2023) Millions of people visit Lake Tahoe each year. Is that sustainable? https://knpr.org/show/knprs-state-of-nevada/2023-02-15/is-the-increased-tourism-to-lake-tahoe-sustainable (archived at https://perma.cc/3CXX-JUV7)

Butvin, S (2023) *Cultural Sustainability in Practice: Community voices in tourism*, Smithsonian Center for Folklife and Cultural Heritage, Washington DC

Calder, S (2025) Venice daytripper entry fee: How much does it cost and how does it work? The Independent. 16 April, www.independent.co.uk/travel/news-and-advice/venice-day-trip-charge-tax-fee-b2734433.html (archived at https://perma.cc/8E3G-QBJ4)

Carroll, A B (2021) Corporate social responsibility: Evolution of a definitional construct, *Business & Society*, **60** (5), pp 922–40

Convention on Biological Diversity (.n.d.) Traditional Knowledge and the Convention on Biological Diversity, www.cbd.int/traditional/intro.shtml (archived at https://perma.cc/M8XY-D8QA)

ECTN (.n.d.) Sustainable Cultural Tourism, www.culturaltourism-network.eu/definitions.html (archived at https://perma.cc/9YEW-ZMAY)

Freeman, R E and Dmytriyev, S (2017) Corporate social responsibility and stakeholder theory: Learning from each other, *Symphonya: Emerging Issues in Management*, 1, pp 7–15, https://doi.org/10.4468/2017.1.02freeman.dmytriyev (archived at https://perma.cc/RA6S-U9SR)

Freeman, R E and McVea, J (2001) A stakeholder approach to strategic management, Darden Graduate School of Business Administration, University of Virginia, Working Paper No. 01-02, https://ssrn.com/abstract=263511 (archived at https://perma.cc/V6C5-GZW4)

Gertner, D (2019) *Tourism, Culture and Identity in Hawaii: Between commercialisation and representation*, Routledge, New York

Hollinshead, K (2018) Rethinking Cultural Representation in Tourism, *Tourist Studies*, 18 (6), pp 649–67

Hosagrahar, J (2023) Culture: at the heart of the Sustainable Development Goals, 11 April, https://courier.unesco.org/en/articles/culture-heart-sustainable-development-goals (archived at https://perma.cc/RYF9-PN6E)

IceHotel (.n.d.) Sustainability practices at the IceHotel, www.icehotel.com/sustainability (archived at https://perma.cc/4DGH-4QG5)

Jones, M (2023) *Designing Sustainable Destinations: Cultural identity and place-making*, Channel View Publications, Bristol

Kiruna Lappland Economic Association (2022) Sustainable Kiruna, https://kirunalapland.se/en/sustainable-kiruna/ (archived at https://perma.cc/ZL4V-HYH8)

Koens, K, Postma, A and Mak, A (2018) Is overtourism overused? Understanding the impact of tourism in a city context, *Sustainability*, 10 (12), p 4384, www.mdpi.com/2071-1050/10/12/4384 (archived at https://perma.cc/3NY7-2HBY)

Lake Tahoe Basin Management Unit (2009) www.fs.usda.gov/r05/laketahoebasin (archived at https://perma.cc/4UMF-2UXD)

Lovato, K (2023) The monster that feeds and eats away Lake Tahoe, CNN Travel, 7 September, https://edition.cnn.com/travel/lake-tahoe-tourism-struggles (archived at https://perma.cc/BF3S-UF5S)

McCool, S F and Bosak, K (2021) *A Research Agenda for Sustainable Tourism*, Edward Elgar Publishing, Cheltenham

McGrath, S and Jonker, J (2023) *Corporate Social Responsibility in THE Industries*, Routledge, London

McKinsey & Company and WTTC (2017) Coping with success: Managing overcrowding in tourism destinations, 14 December, www.mckinsey.com/industries/travel/our-insights/coping-with-success-managing-overcrowding-in-tourism-destinations (archived at https://perma.cc/33WZ-WSAS)

Milano, C (2017) Overtourism and tourismphobia: Global trends and local contexts, www.researchgate.net/publication/323174488_Overtourism_and_Tourismphobia_Global_trends_and_local_contexts (archived at https://perma.cc/68AM-PA9U)

Orefice, C and Nyarko, N (2021) Sustainable value creation in event ecosystems – a business models perspective, *Journal of Sustainable Tourism*, 29 (11–12), pp 1932–47, https://doi.org/10.1080/09669582.2020.1843045 (archived at https://perma.cc/ZV93-LFZC)

Our Homeland Ghana (2023) The art of Kente weaving in Ghana, 30 August, www.ourhomelandghana.com/arts-lifestyle/the-art-of-kente-weaving-in-ghana (archived at https://perma.cc/86VZ-83SR)

Pajon, C (2024) Tradition and innovation in the Japanese tea ceremony, *International Journal of Intangible Heritage*, **19** (2), pp 84–97

Pechlaner, H, Innerhofer, E and Erschbamer, G (2022) *Overtourism: Tourism management and solutions*, Routledge, Abingdon

Pham, K, Andereck, K L and Vogt, C A (2024) Overtourism: A potential outcome in contemporary tourism – Causes, solutions, and management challenges. In D Chhabra, N Atal and A Maheshwari (eds.) *Sustainable Development and Resilience of Tourism*, Springer, Cham

Smith, M and Mair, J (2021) *Events and Sustainability: Managing environmental and social impacts*, Sage, London

UNESCO (2001) Universal Declaration on Cultural Diversity, www.unesco.org/en/legal-affairs/unesco-universal-declaration-cultural-diversity (archived at https://perma.cc/9NJW-ZBWM)

UNESCO (2003) Convention for the Safeguarding of the Intangible Cultural Heritage, www.unesco.org/en/legal-affairs/convention-safeguarding-intangible-cultural-heritage (archived at https://perma.cc/E2PZ-65P2)

UNESCO (2005) Convention on the Protection and Promotion of the Diversity of Cultural Expressions, www.unesco.org/creativity/en/2005-convention (archived at https://perma.cc/BH38-4XF7)

UNESCO (2013) Decision of the Intergovernmental Committee: 8.COM.8.10, https://ich.unesco.org/en/decisions/8.COM/8.10 (archived at https://perma.cc/H3NQ-EYQY)

UNESCO (.n.d.) World Heritage, www.unesco.org/en/world-heritage (archived at https://perma.cc/5AGP-BC2F)

UNESCO Institute for Statistics (2009) Glossary, https://uis.unesco.org/en/glossary (archived at https://perma.cc/9BQM-M5TL)

UNWTO (2011) Tourism Towards 2030: Global Overview, www.globalwellnesssummit.com/wp-content/uploads/Industry-Research/Global/2011_UNWTO_Tourism_Towards_2030.pdf (archived at https://perma.cc/B8EX-D42Z)

UN News (2017) World could see 1.8 billion tourists by 2030 – UN agency, 27 December, https://news.un.org/en/story/2017/12/640512-world-could-see-18-billion-tourists-2030-un-agency (archived at https://perma.cc/7MNL-5Z39)

UN Tourism (.n.d.) www.unwto.org/ (archived at https://perma.cc/APT7-BNCC)

UNWTO (2018) Overtourism? Understanding and managing urban tourism growth beyond perceptions, September, www.e-unwto.org/doi/book/10.18111/9789284420070 (archived at https://perma.cc/PJ9N-JJW7)

US Census Bureau (2020) South Lake Tahoe city, DEC demographic and housing characteristics, https://data.census.gov/table/DECENNIALDHC2020.P1?q=P1&g=010XX00US_040XX00US06_160XX00US0673108&d=DEC+Demographic+and+Housing+Characteristics

PART TWO
Applying sustainability in professional practice

5 | Measuring impact and reporting

| Systems thinking competency | Critical thinking competency | Strategic competency | Normative competency |

CHAPTER AIM

The aim of this chapter is to provide a clearer understanding of concepts such as ethical businesses, corporate social responsibility, environmental and social governance, and Scope 1/2/3 emissions. It provides guidance on ethical compliance and sustainable practices for a more responsible business. Emphasis is placed on aligning impact measurement and reporting practices with key global frameworks, including the Sustainable Development Goals (SDGs), specifically SDG 12, SDG 13, and SDG 16.

The chapter supports the development of key sustainability competencies as identified by UNESCO (2017): systems and critical thinking, strategic competency, and normative competency. Through real-world examples, project-based learning and reflective practice, students will explore how impact measurement and reporting can be embedded into business operations, risk management and sustainability auditing. The chapter encourages learners to think critically, question prevailing practices and identify innovative solutions for a more sustainable and responsible business environment.

LEARNING OUTCOMES

Upon completion of this chapter, you will be able to:

- identify key concepts of corporate social responsibility
- recognize effective ESG implementation
- critically evaluate measuring and sustainability reporting
- understand the role of international certifications in promoting responsible business practices
- apply practical skills in sustainability auditing
- develop critical thinking for sustainability problem-solving

KEY WORDS

Corporate social responsibility (CSR), environmental, social and governance (ESG), ethical business, sustainability reporting, Scope 1/2/3 emissions, international certifications.

Introduction

This chapter aims to provide a comprehensive understanding of how to assess and report the environmental, social and economic impacts of tourism, hospitality and events (THE) businesses. It covers essential concepts such as ethical business practices, corporate social responsibility (CSR) and environmental, social and governance (ESG). CSR is a broader, more qualitative concept that emphasizes ethical business practices and a positive impact on society and the environment. ESG is a more structured, quantitative framework that focuses on specific, measurable environmental, social and governance factors to reduce risk and increase sustainability (see Figure 5.1 below). The chapter also discusses the importance of international certifications, the categorization of emissions into Scope 1, 2 and 3, and offers practical tips for reducing carbon footprints.

By systematically measuring and reporting sustainability metrics, THE businesses can enhance transparency, ensure accountability, manage risks and engage stakeholders effectively. In conclusion, the chapter encourages students to put into practice some of these solutions through an activity.

In the THE sector, sustainability is a shared responsibility that spans across various roles and functions. Depending on their position, individuals may have the responsibility to make decisions that impact sustainability outcomes. For instance, a chef might be in charge of sourcing sustainable ingredients and managing food waste, while an operations manager might oversee energy efficiency and property maintenance. Each role, whether directly or indirectly, contributes to the overall sustainability goals of the organization. As future professionals in THE, it is important to recognize that while one person may not be responsible for all actions related to Scope 1, 2 and 3 emissions, every job has a role to play. Understanding how your specific responsibilities can influence sustainability practices will enable you to make informed decisions that support a more sustainable and resilient industry.

Measuring impact and sustainability reporting are essential practices within the THE industry. These practices enable businesses to understand and communicate

Figure 5.1 CSR vs ESG differences

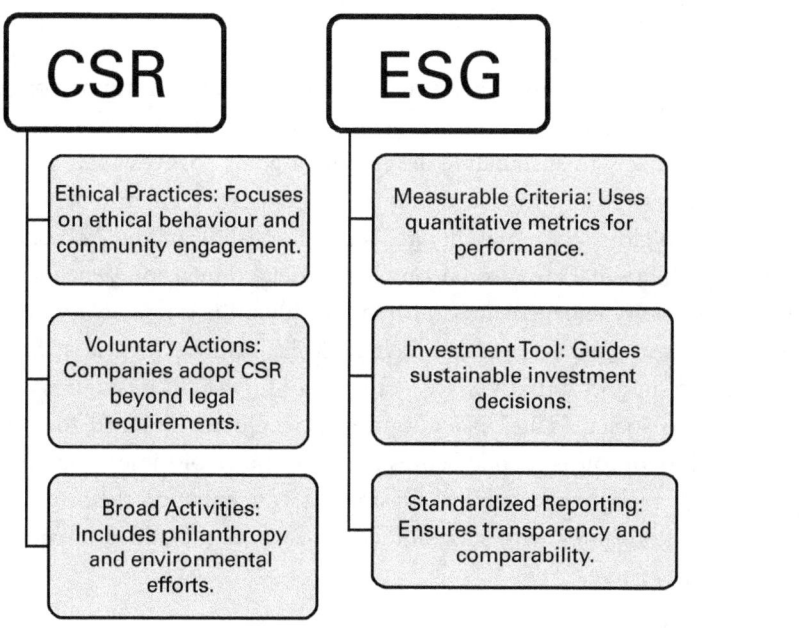

their ESG performance. By systematically assessing their impact, THE businesses can identify areas for improvement, demonstrate their commitment to sustainability and enhance their transparency and accountability. This process involves collecting, analyzing and reporting data on various sustainability metrics, which helps businesses align their operations with broader sustainability goals. The primary goals of sustainability reporting include enhancing transparency, ensuring accountability, managing risks and engaging stakeholders. These practices are essential for building trust with stakeholders, including customers, investors and regulators.

By clearly communicating their ESG performance, businesses can foster a culture of accountability and proactive management, driving continuous improvement in their sustainability efforts. Sustainability reporting also helps businesses identify potential risks and opportunities related to their environmental and social impacts, enabling them to make informed decisions and improve their overall performance (Luque-Vílchez et al, 2023).

The THE industry has a significant environmental footprint, encompassing energy consumption, water usage, waste generation and carbon emissions. Sustainable practices are crucial to mitigate these impacts. For instance, implementing energy-efficient technologies and waste reduction programmes can substantially reduce the environmental burden of THE operations. The Statistical Framework for Measuring the Sustainability of Tourism (SF-MST) is an internationally agreed framework designed to integrate and organize statistics related to the economic, environmental and social aspects of sustainable tourism (UNWTO, 2024). It aims to provide a comprehensive approach to understanding and measuring how tourism impacts and depends on these three dimensions.

The SF-MST (see Figure 5.2 below) helps countries produce reliable, comparable data that can be used to assess the sustainability of tourism activities and their contributions to the economy, society and the environment. This framework can support more effective decision-making and policy development to ensure that tourism contributes positively to sustainable development goals (UNWTO, 2017). By adopting these practices, businesses can minimize their negative environmental impacts and contribute to global sustainability goals (Anjum, Ghufran and Abbas, 2024).

Sustainability in THE also encompasses social dimensions, such as labour practices, community engagement, and human rights. Businesses play a pivotal role in promoting social equity and wellbeing. Ethical labour practices, fair wages and community support initiatives are examples of how THE businesses can contribute positively to society. The UN's adoption of a global standard for measuring the sustainability of tourism underscores the importance of these social dimensions. By prioritizing social responsibility, businesses can enhance their reputation, build stronger relationships with stakeholders and create a positive impact on the communities they serve.

Figure 5.2 The Statistical Framework for Measuring the Sustainability of Tourism (SF-MST)

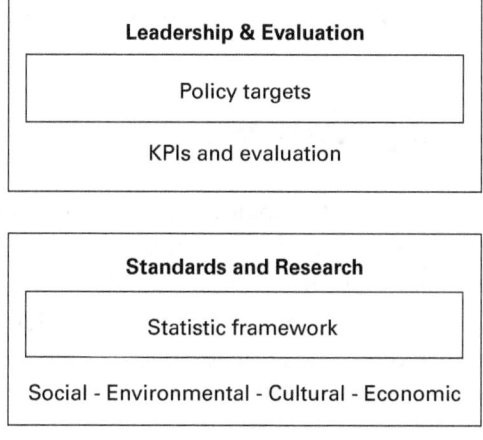

Adapted from UNWTO, 2024

Integrating sustainability into business operations can also drive economic benefits. These include cost savings from efficient resource use, enhanced brand reputation, and increased customer loyalty. For example, hotels that adopt green certifications often attract eco-conscious travellers, leading to higher occupancy rates and customer satisfaction. By demonstrating their commitment to sustainability, businesses can differentiate themselves in the market, attract investment and achieve long-term financial success.

The ESG Framework for Tourism Businesses, developed by UN Tourism, is designed to help tourism enterprises measure and report their sustainability efforts consistently and reliably. ESG covers environmental, social and governance, which are key non-financial factors that businesses need to consider ensuring sustainable operations (UNDP, 2024).

The tourism industry faces growing regulatory pressures and market demands for sustainability. Key regulations such as the Corporate Sustainability Reporting Directive (CSRD) and the Task Force on Climate-related Financial Disclosures (TCFD) are shaping the landscape (Operato et al, 2025). The CSRD is an EU directive that enhances and modernizes the rules on social and environmental reporting for large companies and listed SMEs,. It aims to improve the quality, comparability and reliability of sustainability information disclosed by companies. The TCFD, established by the Financial Stability Board (FSB), aimed to improve and increase the reporting of climate-related financial information. Although it has now disbanded, its recommendations continue to be influential and are now monitored by the IFRS Foundation.

Additionally, there is an increasing consumer preference for eco-friendly products and services, which businesses must address to remain competitive. The adoption of the Statistical Framework for Measuring the Sustainability of Tourism by the UN highlights the global commitment to standardized sustainability reporting. By staying ahead of regulatory requirements and market trends, businesses can ensure compliance, enhance their competitive edge and contribute to a more sustainable future.

Ethical business practices

Ethical business practices refer to the application of moral principles and standards in business operations. These practices ensure that businesses operate in a manner that is fair, transparent and respectful to all stakeholders, including employees, customers, suppliers and the broader community. In this section we will explore the definition and importance of ethical business practices, followed by the principles that guide these practices within the THE industry.

Definition and importance of ethical business practices

According to the UN World Tourism Organization's Global Code of Ethics for Tourism (1999), ethical business practices are crucial for several reasons. The following five components explain why:

1 **Trust and reputation**: Ethical practices build trust with customers and other stakeholders. A reputation for integrity can differentiate a business in a competitive market, leading to customer loyalty and long-term success.

2 **Legal compliance**: Adhering to ethical standards often aligns with legal requirements, helping businesses avoid legal issues and penalties. This compliance is particularly important in industries like tourism, hospitality and events, where regulations can be stringent.

3 **Employee satisfaction**: Ethical practices create a positive work environment, leading to higher employee morale and retention. Employees are more likely to be engaged and productive when they feel their employer values fairness and integrity.

4 **Sustainable growth**: Ethical business practices contribute to sustainable growth by fostering a culture of responsibility and long-term thinking. This approach helps businesses manage risks and seize opportunities in a rapidly changing market.

5 **Community relations**: Businesses that operate ethically are more likely to gain the support and goodwill of the communities in which they operate. This support can be crucial for business operations, especially in the tourism and hospitality sectors, which often rely on local resources and attractions.

Having established the importance of ethical business practices, we now turn our attention to the principles that guide these practices within THE.

Principles of ethical business in the THE industry

The principles of ethical business in THE emphasize the importance of balancing economic success with social and environmental responsibility. Key principles include responsible stewardship of the environment, which involves avoiding wasteful use of resources and striving to protect and improve the environment (Ateeq and Milhem, 2024). This can be achieved by reducing emissions, conserving water and minimizing waste, such as through the implementation of energy-saving measures and recycling programmes in hotels. Respect for people and communities is another crucial principle, where ethical businesses consider the views and needs of local communities, ensuring their operations do not harm local cultures or economies. This includes fair treatment of employees, support for local suppliers and engagement in community development projects.

Fairness and transparency are also vital, with businesses operating transparently and clearly communicating their practices and performance to stakeholders. This involves honest marketing, transparent reporting of sustainability efforts and fair treatment of all business partners. Engagement and collaboration with stakeholders, including customers, employees and local communities, are essential for understanding their needs and concerns, leading to innovative solutions and stronger relationships. Finally, ethical businesses commit to continuous improvement in their practices, regularly assessing and reporting on their social and environmental impacts and seeking ways to enhance their positive contribution.

With a clear understanding of ethical business practices, the concept of corporate social responsibility (CSR) and its application in the THE industry can now be further explored.

Corporate social responsibility (CSR)

In this section, we will explore the definition of CSR, its significance in the THE industry, and the key components that make up effective CSR strategies. For Zhang and Hao (2024), CSR in THE refers to the commitment of businesses to contribute to sustainable economic development while improving quality of life for their workforce, their families, the local community and society at large. CSR in the THE industry encompasses a wide range of activities, from environmental conservation efforts to social equity initiatives and ethical business practices.

Understanding CSR in the THE industry

CSR is particularly important in THE due to its significant impact on local communities and environments. Tourism and hospitality businesses often operate in areas of natural beauty or cultural significance, making it essential for them to minimize their environmental footprint and support local communities. Additionally, the industry is labour-intensive, relying heavily on the local workforce, which underscores the importance of fair labour practices and community engagement.

The evolution of CSR in the THE industry has been driven by several factors, including increased consumer awareness, regulatory pressures, and the recognition that sustainable practices can lead to long-term business success. Companies are increasingly adopting CSR strategies to enhance their brand reputation, attract socially conscious consumers and mitigate risks associated with environmental and social issues.

Key components of CSR

CSR in the THE industry encompasses several key components: environmental sustainability, social responsibility, economic responsibility, ethical governance and stakeholder engagement (Aw, Etim and Ogbonda, 2024):

- **Environmental sustainability** focuses on reducing the environmental impact of business operations through practices such as energy efficiency, waste reduction, water conservation and the use of sustainable materials. For instance, hotels may implement green building standards, reduce single-use plastics and promote eco-friendly tourism activities.
- **Social responsibility** involves ensuring fair labour practices, supporting local communities, and promoting social equity by providing fair wages, safe working conditions and opportunities for career development. Businesses may also engage in community development projects, support local education initiatives and promote cultural preservation.
- **Economic responsibility** entails conducting business in a financially sustainable manner while contributing to the local economy by supporting local suppliers, creating job opportunities and ensuring that economic benefits are shared with the community.
- **Ethical governance** involves maintaining high standards of integrity and transparency in business operations, adhering to legal and regulatory requirements, implementing anti-corruption measures, and ensuring accountability through regular reporting and stakeholder engagement.

- Finally, effective CSR requires **active engagement with all stakeholders**, including employees, customers, suppliers and the local community, through regular communication, feedback mechanisms and collaborative initiatives to address shared concerns and goals.

Benefits of CSR for businesses and communities

CSR initiatives play a crucial role in the THE industry (Coles, Fenclova and Dinan, 2013; Ogbeide et al, 2017).

Enhanced reputation is one of the significant benefits, as CSR initiatives can make a company more attractive to consumers, investors and potential employees by highlighting a strong[er] commitment to social and environmental responsibility. This differentiation in a competitive market can lead to increased customer loyalty, as consumers are increasingly seeking out businesses that align with their values. By demonstrating a commitment to CSR, companies can build stronger relationships with their customers and foster loyalty. Additionally, many CSR practices, such as energy efficiency and waste reduction, can lead to cost savings and improved operational efficiency, benefitting both the environment and the bottom line. CSR also helps businesses identify and mitigate risks related to environmental and social issues, allowing them to avoid potential legal issues, reputational damage and operational disruptions. Furthermore, CSR initiatives that support local communities can lead to stronger community relations and a more supportive operating environment, which is particularly important in the THE industry, where businesses often rely on local resources and goodwill.

Examples of CSR initiatives

Here are three examples of popular CSR initiatives from global hotel corporations:

- **Hilton's 'Travel with Purpose'**: Hilton's CSR programme focuses on creating shared value for its business and communities. The initiative includes goals for reducing environmental impact, such as cutting carbon emissions and water consumption, as well as social goals like promoting diversity and inclusion and supporting local communities through volunteerism and donations (Hilton, 2024).
- **Marriott's 'Serve 360'**: Marriott International's CSR strategy, 'Serve 360: Doing Good in Every Direction,' aims to make a positive and sustainable impact wherever it does business. The programme includes initiatives to reduce environmental impact, support local communities and promote ethical business practices. Marriott has set ambitious goals, such as reducing water intensity by 15 per cent and achieving 100 per cent of its hotels certified to a recognized sustainability standard (Marriott, .n.d.).

- **Accor's 'Planet 21'**: Accor's Planet 21 programme focuses on sustainable development and includes initiatives to reduce energy and water consumption, promote sustainable food practices and support local communities. The programme also emphasizes employee engagement and training in sustainability practices (Accor, 2021).

Having reviewed key aspects of CSR, environmental, social and governance (ESG), a framework that provides tangible metrics for assessing and managing sustainability issues, will be the topic of the next section.

Introduction to Environment, Social and Governance (ESG)

In the business environment, sustainability, social responsibility and good governance are becoming increasingly important. The concept of ESG has become a key framework for assessing and managing these issues (UNDP, 2024). This section will provide an introduction to ESG, including its historical context, criteria and relevance to the THE industry. We will also discuss the benefits of ESG reporting, risk management and stakeholder engagement. Associated with CSR, ESG provides more tangible metrics. If CSR focuses on the result, ESG is increasingly used as a performance measure to inform decision-making and risk. ESG is often integrated into company strategy to demonstrate responsibility.

According to the UNDP (2024):

- E is for Environmental: it assesses the impact of a company on climate change, carbon emission, pollution, waste management and resource use.
- S is for Social: it assesses how a company impacts society around them. This focuses on relationships with their teams, suppliers, customers and the local community. It also includes diversity, human rights and labour.
- G is for Governance:, it assesses a company's transparency and ethics, including executive pay, shareholders' rights, compliance with industry standards and board composition.

Historical context

ESG started in the 1950s as Socially Responsible Investing (SRI), which allowed investors to avoid investing in companies that did not align with their values, such as tobacco or armaments (Townsend, 2020). By the 1980s, CSR had gained popularity, and investors started incorporating social and environmental impacts into their decision-making. By early 2000, ESG was first referenced in a UN Global Compact Partnership with a major financial institution in a report called 'Who Cares Wins', and in 2005 the UN launched the Principles of Responsible Investing (PRI) which encouraged the incorporation of ESG factors into investment decisions.

From 2010 onwards, ESG was widely adopted for its reputational and financial benefits, leading to further ratings and frameworks such as TCFD (Task Force on Climate-Related Disclosures). Since 2020, adopting and reporting against ESG frameworks have dramatically increased as climate change, inequality and corporate transparency have become increasingly important for stakeholders and investors (Townsend, 2020).

ESG criteria and their relevance to the THE industry

ESG reporting provides tangible metrics for transparency, accountability and continuous improvement, helping businesses manage risks and attract eco-conscious consumers. Therefore, ESG should be incorporated into THE businesses as these not only consume large amounts of energy, water and resources associated with food and drink, but they are also highly dependent on people, both customers and employees (Hasan et al, 2024). By having a strong social framework, they are in a more competitive position to retain staff through higher levels of satisfaction.

Potential customers are also increasingly researching a business's sustainability credentials before deciding who to choose; having a robust ESG framework puts a business in a competitive position. 50 per cent of consumers said sustainability is one of their top four key purchase criteria when purchasing products or services. By its very nature, the THE industry depends on reputation, so a strong environmental and social framework backed up by a transparent governance strategy can increase the confidence of stakeholders, particularly investors (Hasan et al, 2024).

Benefits of ESG reporting

Although reporting ESG can be demanding, there are significant benefits to it. By using ESG reporting frameworks, a business can provide and use tangible metrics to create a transparent view of their company's impact, making them accountable for monitoring progress and improvement, from carbon emissions to gender diversity rates. Significantly, it also reduces the opportunity for greenwashing accusations (see Chapter 9), as any claims can be substantiated through data, and allows businesses to identify future risks such as supply chain issues. Using the ESG strategy can also provide a competitive advantage and attract increasingly eco-conscious consumers.

Risk management

Understanding and mitigating risks related to environmental and social impacts is critical for THE businesses. The industry depends on supply chains, and with rapid climate change, the likelihood of these being disrupted is high, so understanding their resilience is critical. The coffee industry heavily relies on supply chains, and rapid climate change poses significant disruption risks. For instance, severe frost in Brazil's coffee belt in June and July 2021 caused global coffee prices to spike by 13 per cent.

Additionally, changing climatic conditions are expected to render some coffee-growing regions unsuitable, leading to the emergence of new pests and diseases, and loss of productive lands, which will invariably affect coffee farming and supply chains. Understanding and enhancing the resilience of these supply chains is therefore critical. (Bermudez et al, 2022). Based on global climate models used to explore three climate scenarios and soil conditions, highly and moderately suitable Arabica coffee-growing environments around the world are expected to decrease by 50 per cent and 30 per cent respectively by 2050 (Grüter et al, 2022).

A business can mitigate the effects by understanding which products are at risk. Recognizing the social impact of products and services will also identify vulnerabilities, such as modern-day slavery practices, and can prompt action. This is also the case for environmental risks, such as energy price increases, which can again be mitigated to a certain extent through implementing energy management procedures without assessing these issues. Finally, with the likelihood of rising sea levels and increasingly extreme weather, ESG can identify if any properties are at risk and put contingency plans in place. The above improves the business's reputation and prepares the company for future regulations.

Stakeholder engagement

Engaging stakeholders in ESG reporting aligns values and expectations, enhancing reputation and loyalty. Stakeholders must be engaged with ESG reporting and the associated actions as it aligns values and expectations from employees, customers and the local community to investors. 60 per cent of companies also report substantial pressure from their board members and management teams (Steinmann, 2024). A business can strengthen its reputation and employee and customer loyalty by demonstrating transparency in its ESG reporting and regularly updating the public on this. A Deloitte survey reveals growing climate change awareness, with 64 per cent of consumers changing their purchase habits to address climate change, up from 53 per cent the previous year (Steinmann, 2024).

Steps to implement ESG in the THE Industry

To remain resilient and competitive in a changing context, a thorough assessment and strategy are required. The exact depth of evaluation will be determined by organizational size and the desired level of granularity. Nevertheless, the following broad steps are to be undertaken.

1 Assessment and benchmarking

   Current performance should be analysed to identify strengths, weaknesses, risks and opportunities.

   Where possible, performance should be benchmarked against relevant industry peers.

2 Materiality and focus

Key ESG areas most relevant to stakeholders and the business are to be identified and prioritized.

3 Goal-setting and KPIs

Clear, measurable and attainable goals are to be established once the above are confirmed.

Short-, medium-, and long-term targets, supported by practical KPIs, should be defined to ensure realism.

Overextension is to be avoided; focusing on two or three priority strategies is recommended rather than dispersing resources too widely.

4 Framework alignment

Goals should be aligned to an external framework, such as the UN SDGs.

5 Strategy and roadmap development

Assessment results are to be used to develop an ESG strategy and an aligned roadmap that:
- outlines key initiatives (e.g. energy reduction, supply-chain resilience)
- sets a timeline with ambitious yet achievable milestones
- allocates resources, budgets and responsibilities, with a single accountable lead for each initiative

6 Implementation

Initiatives are to be implemented according to the plan. Careful sequencing should be applied, as excessive simultaneous change can cause rollout failures, particularly for internal changes.

Messaging should be kept clear and straightforward and, where possible, integrated into existing procedures to maximize adoption.

7 Monitoring, reporting and improvement

Once implemented, initiatives are to be monitored and reported on regularly to communicate progress, with the expectation that continuous review and improvement will be required.

Reporting should be written and presented for accessibility, considering plain language and ease of access.

Challenges should be reported honestly while successes are highlighted and celebrated, which can enhance brand reputation and strengthen employee and stakeholder pride (Fatima and Elbanna, 2023).

8 Measurement and audit

The impact of initiatives is to be measured and audited, with particular attention to carbon footprint and sustainability assurance.

Carbon footprint and Scope 1, 2 and 3 emissions

Implementing sustainability initiatives requires measuring and auditing their impact. Two critical tools for this are carbon footprint measurement and sustainability auditing. In this section, we will focus on understanding and categorizing emissions into Scope 1, 2 and 3, and provide strategies for reducing these emissions to enhance sustainability performance. These tools help businesses understand their environmental impact and identify improvement opportunities. Measuring carbon footprints highlights areas for targeted emission reductions. The tourism sector, responsible for about 5 per cent of global greenhouse gas emissions, urgently needs sustainability measures (Cambridge Institute for Sustainability Leadership, 2024). Sustainability audits, focusing on carbon footprints, can be conducted internally or with external partners for detailed analysis. Identifying high carbon impact areas allows management to take targeted actions to reduce emissions and enhance sustainability performance.

A critical step in understanding and managing a business's carbon footprint is categorizing emissions into three distinct scopes, as defined by the Greenhouse Gas (GHG) Protocol (GHG Protocol, .n.d.) (see below). This categorization provides a comprehensive framework for assessing a company's total carbon footprint and helps distinguish between direct and indirect emissions. By doing so, businesses can better identify, manage and mitigate their sources of greenhouse gas emissions. This structured approach also supports the setting of accurate and achievable reduction targets, ensuring that all aspects of a business's operations are considered in sustainability planning (Climate Change Committee, 2020).

Scope 1 emissions are direct emissions from fuel combustion for transportation or heating. Scope 2 emissions are indirect emissions from purchased electricity, heating and cooling. Scope 3 emissions include all other indirect emissions in the value chain, such as guest travel, product life cycle and waste management. See Figure 5.3.

Figure 5.3 Scope 1, 2 and 3 emissions

Scope 1,2 and 3 EMISSIONS		
Scope 1 Combustion of fuels	Scope 2 Electricity and heating	Scope 3 Transportation and waste

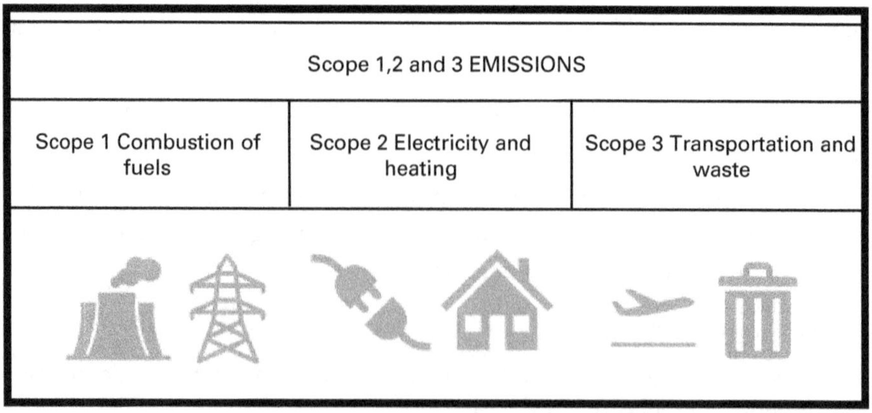

Scope 1: Direct emissions

Scope 1 emissions are direct emissions from sources owned or controlled by a company, including refrigerant leaks, fuel from vehicle fleets and emissions from gas-powered cooking appliances.

To reduce these emissions, businesses can take several steps, such as managing refrigerants. Analyze the refrigerants used, focusing on low Global Warming Potential (GWP) options, especially F-gases, while ensuring that refrigerant leaks are kept to a minimum through regular maintenance.

Additionally, a vehicle fleet can be optimized using route optimization software, by ensuring journeys are made only when necessary, and prioritizing when possible night-time deliveries. Choosing online meetings over in-person ones can also reduce emissions. Also, there should be consideration given to gradually replacing petrol vehicles with electric ones as required.

Scope 2: Indirect emissions from energy consumption

Scope 2 addresses a business's energy usage, particularly electricity and gas, highlighting cost savings and carbon emission reduction opportunities. While these emissions are indirect, they represent significant opportunities for cost savings and carbon reduction.

Businesses can engage their teams and achieve significant savings by implementing behavioural changes – such as turning off equipment overnight and optimizing its use during trading hours. Promoting efficient equipment operation and adopting low-carbon technologies like voltage optimizers and light timers can also enhance these efforts.

For businesses using renewable energy with Renewable Energy Guarantees of Origin (REGO), Scope 2 emissions are low, but reducing energy consumption further is still essential to minimize grid strain and maximize financial benefits.

Scope 3: Indirect emissions from the value chain

Scope 3 emissions encompass all other indirect emissions that occur in a company's value chain. These emissions often represent the largest portion of a business's carbon footprint and include activities such as supplier operations, waste management and water usage. Scope 3 is divided into 15 categories, making it the most complex scope to measure and manage.

Suppliers and producers

When selecting new suppliers and products, businesses should prioritize those with a measurable carbon footprint and ESG goals that align with their own sustainability objectives. For existing suppliers and producers who have not yet measured their carbon footprint, it is essential to encourage them to do so. Additionally, some suppliers and producers can provide a life cycle assessment (LCA) for critical products, which offers a detailed assessment of the environmental impact across the product's life cycle. This information can significantly support the business's carbon footprint calculations and ESG planning.

Waste

While waste accounts for a smaller proportion of Scope 3 emissions, its reduction is a critical component of achieving ESG goals. The first step is to identify the primary sources of waste within your operations. In the hospitality industry, food waste is often a significant contributor. Conducting a food waste audit, such as the Guardians of Grub programme developed by WRAP (Waste and Resources Action Programme), and following a structured food waste reduction roadmap can yield substantial benefits. For more information on food waste reduction strategies see Chapter 7; for a food waste reduction road map refer to WRAP's Courtauld Commitment 2030 Annual Report (WRAP, 2024). The Courtauld Commitment is currently known as the UK Food and Drink Pact (for more information please refer to Further Resources).

Waste can be reduced by implementing efficient recycling training and ensuring that bins are conveniently placed to encourage recycling and further reduce waste. Prioritizing minimal and recyclable or compostable packaging is also essential and can be assessed through a packaging audit. Managing equipment disposal by properly handling the disposal of equipment, such as refrigerators and Waste Electrical and Electronic Equipment (WEEE), to ensure environmentally responsible practices.

Water

Although water usage has a relatively small carbon footprint, it remains an essential element of a sustainability audit. Efficient water management can uncover significant opportunities for operational savings and environmental impact reduction. For example, fixing a dripping hot tap can save approximately 20 litres of water per day and around 8kWh of energy used to heat it (contributing to Scope 2 emissions) (Molinos-Senante et al, 2022). Businesses should regularly monitor water usage, identify inefficiencies, and implement water-saving measures to enhance sustainability performance.

Stakeholders

Engaging stakeholders – particularly employees and guests – is crucial to the success of any carbon reduction initiative. Transparency is key: openly communicate both successes and challenges to build trust and foster a sense of shared responsibility. This approach not only strengthens stakeholder buy-in but also mitigates the risk of greenwashing. By involving stakeholders in any sustainability efforts, the business can create a collaborative environment that supports long-term progress toward achieving the ESG goals.

Beyond the established framework of Scopes 1, 2 and 3, it is important for future professionals to be aware of an emerging concept in sustainability debates: Scope 4, or 'avoided emissions'. Unlike the first three scopes, which measure the emissions a company produces, Scope 4 is a voluntary metric that aims to quantify the positive climate impact of a business. TEAM (.n.d.) calculates the emissions reductions that occur because a customer uses a company's product or service instead of a more carbon-intensive alternative. While this concept offers a competitive advantage by showcasing a product's environmental benefits, it is not yet officially recognized or fully standardized and does not appear in standard diagrams like Figure 5.3. Experts caution that any Scope 4 claims must be reported transparently and separately from Scopes 1, 2 and 3 to avoid accusations of greenwashing, as they should not be used to offset a company's own carbon footprint (TEAM, .n.d.).

With a clearer understanding of emissions and their nature, in the next section the role of sustainability auditing is explored as a practical tool to identify best practices or inefficiencies.

Introduction to sustainability audits

Auditing for sustainability in the tourism sector is not merely a compliance exercise but a strategic tool for risk management and enhancing competitive advantage. This section will introduce the concept of sustainability audits, outline their objectives and detail the key components and steps involved in conducting a thorough audit. Sustainability audits are a powerful mechanism for promoting responsible business practices, enhancing the guest experience, creating a competitive edge and engaging employees. By fostering a culture of accountability and transparency, audits can cultivate an engaged and loyal workforce.

Contrary to the common misconception that sustainability initiatives are costly and offer limited return on investment (ROI), reducing resource use through operational efficiency can significantly improve profitability. Research indicates barriers to sustainability adoption in hospitality include high perceived costs and doubts about ROI, while motivators include cost savings, competitive advantage and improved public image (Abul Basher Rasel, 2024). Sustainability audits provide an excellent opportunity to identify inefficiencies, such as outdated HVAC (heating, ventilation and air conditioning) systems or excessive lighting use, enabling businesses to cut carbon emissions and operational costs simultaneously.

For example, Hilton's Green Breakfast initiative, which focused on measuring food waste, identifying trends, and coaching kitchen teams, reduced food waste by

62 per cent in pilot hotels (Hilton, 2024). This initiative not only demonstrated scalable solutions to sustainability challenges, but also improved profitability by reducing waste and increasing operational efficiencies.

Beyond immediate cost savings, sustainability audits prepare businesses for future challenges, from compliance with evolving regulations to mitigating environmental and social risks in the supply chain. Audit outcomes can guide procurement teams to source more sustainable products, such as local produce, eco-friendly materials or upcycled fixtures and fittings. Additionally, audits establish a baseline for annual reporting, demonstrating transparency and progress. Many businesses align their audit outcomes with certifications, such as ISO 14001, LEED, Green Key, B Corp and EarthCheck (see 'International certifications' later in this chapter) to enhance their credibility and market position.

Objectives of a sustainability audit

A sustainability audit does not have to follow a specific framework but must be thorough. To do this, objectives should be considered, and questions should be developed to cover the following.

Compliance and governance

Accurate and detailed data is essential for a business to meet current and upcoming legislation and industry standards without requiring excessive additional work to supply this documentation. It also provides tangible baselines, goals and KPIs against which to track progress.

Enhancement of overall sustainability performance

Completing a thorough audit can identify areas of waste and efficiency opportunities, identifying areas of resource waste while improving operational efficiencies and creating a competitive edge. Transparent audit reporting will also engage employees and guests with broader ESG goals.

Risk management

Environmental, social, operational or reputational risks must be considered at all stages of an audit to prevent them from becoming more considerable risks in the future.

Key components of sustainability audits

Since 80 per cent of the THE industry's carbon footprint comes from Scope 3 emissions, businesses must collaborate with and support their suppliers in their sustainability journey. This collective effort is critical to reducing the industry's environmental impact.

Suppliers

A vital element is a supplier risk assessment, which should focus mainly on their products' environmental and social impact.

Environmental

In addition to assessing the immediate environmental impact of the business, the next focus should be on suppliers and producers. Utilizing their environmental impact assessment will ensure that ecological practices, such as water management and energy use, are monitored and reduced. Completing this assessment can also highlight supply-chain risks, such as ethical conduct or political instability, before they become damaging.

Social

Completing a social responsibility policy with suppliers will provide confidence or identify risks in labour practices, such as fair wages, safe working conditions and workers' rights. Most suppliers should have an audit demonstrating the above, so in these cases your responsibility is to ensure any high-risk areas have been followed up on and progress tracked. Social audits, as part of sustainability assessments, can help the hospitality industry address labour rights and ensure fair practices within the supply chain. Completing or viewing existing environmental and social responsibility audits alongside your supplier provides an opportunity to encourage suppliers to implement further sustainable practices while improving your relationship with them.

Governance

This is fundamental to ensuring sustainability implications are considered in every business decision, fully integrating them into day-to-day operations and incorporating them into business planning. It allows a business to demonstrate its commitment and accountability to all stakeholders. Through solid governance from both your business and the suppliers, you can be confident that your sustainability practices are upheld, producing long-term resilience and trust.

Steps to conduct a sustainability audit: A five-step approach

Sustainability audits can be conducted effectively by following a structured five-step approach (see Figure 5.4 below). This method ensures a thorough and systematic process, enabling businesses to identify risks, uncover opportunities, and achieve their ESG goals.

Figure 5.4 A five-step approach to sustainability auditing

1 Planning
- Define the scope, goals and criteria to avoid scope creep.
- Decide whether to focus on specific departments (e.g. food and beverage) or broader aspects (e.g. waste and energy).
- Identify and prepare key stakeholders and suppliers for the audit process.
- If a carbon footprint assessment exists, prioritize auditing suppliers with the highest emissions.
- If no assessment exists, start with major suppliers (e.g. dairy/meat providers) for a manageable and impactful scope.

2 Gathering Data
- Collect qualitative and quantitative data from internal and external stakeholders.
- Obtain internal data (e.g. energy bills), noting potential gaps due to inadequate systems.
- Conduct interviews with key stakeholders to fill data gaps.
- Use a supplier engagement questionnaire to evaluate sustainability practices.
- Ensure supplier data sets are comparable.
- Initial data collection is time-intensive but becomes more efficient as systems improve.

3 Data Analysis
- Analyze data to identify historical trends, establish baselines, and forecast future usage.
- Examine seasonal variations in energy consumption to understand patterns and inefficiencies.
- Highlight risks, such as non-compliant practices or energy-intensive processes.
- Conduct additional risk assessments on high-impact areas, like supply chain vulnerabilities.
- Identify opportunities for improvement to drive sustainability and operational efficiency.

4 Insights and Actions
- Translate analyzed data into actionable insights and recommendations aligned with ESG goals, resources and departmental capabilities.
- Categorize actions into short-, medium- and long-term initiatives, detailing their impact, financial support and feasibility.
- Establish measurable goals to create a clear roadmap for achieving ESG objectives
- Track progress over time to ensure continuous improvement and alignment with sustainability targets.

5 Follow Up
- Conduct regular follow-ups and annual audits to ensure continuous improvement.
- Revise goals, timelines and KPIs as needed to reflect changing business circumstances.
- Monitor progress to ensure actions deliver desired outcomes and maintain sustainability as a priority.
- Sustain focus on sustainability even during operational pressures like labour shortages or sales targets.
- Demonstrate governance, maintain compliance and reinforce the business's commitment to sustainability.

Having outlined the steps for a sustainability audit, and therefore potentially identifying areas for improvement, the next section provides practical tips for reducing carbon footprints.

Industry tips for reducing carbon footprint

> **INDUSTRY VOICE: KAT EMMETT**
>
> Kat Emmett, a sustainability expert with 25 years of experience in the industry, emphasizes, 'Key learning is to resist the urge to change everything at once. Instead, concentrate on one aspect at a time, make incremental adjustments, and integrate new behaviours into existing practices while measuring their impact. This incremental approach, combined with robust measurement and reporting, is crucial for achieving lasting and meaningful change. Furthermore, engaging with industry initiatives like Roadmap to Net Zero or the Net Zero Carbon Events Roadmap provides valuable resources and guidance for businesses seeking to align their sustainability efforts with broader industry goals. The roadmap offers a company pathway to net zero and describes priority action areas, including powering events efficiently with clean energy and reducing waste.'
>
> (See Zero Carbon Forum, 2022; Net Zero Carbon Events, 2022.)

For businesses embarking on their sustainability journey, identifying and addressing 'low-hanging fruit', simple changes requiring minimal investment yet yielding significant impact, is crucial (Kat Emmett, see Industry Voice box above). These initial successes not only demonstrate the feasibility of sustainable practices but also foster team buy-in and create momentum for more ambitious initiatives. Focusing on energy reduction, particularly electricity consumption, offers a prime opportunity for this, as it directly translates to lower carbon emissions, cost savings and heightened awareness among all levels of the business. This section will explore the components of a carbon footprint and provide actionable strategies for reduction in the short, medium and long term, emphasizing behavioural change, supplier engagement, maintenance optimization and a revaluation of established practices. These strategies will contribute to reducing both the carbon footprint and broader ESG impacts (Ateeq and Milhem, 2024).

Understanding Scope 1, 2 and 3 emissions

As discussed earlier in this chapter, understanding the different scopes of emissions is fundamental to developing a comprehensive carbon reduction strategy. Scope 1 emissions are direct emissions from owned or controlled sources, such as on-site fuel combustion for heating or company-owned vehicles. Scope 2 emissions are indirect emissions from the generation of purchased energy, primarily electricity. Scope 3 emissions encompass all other indirect emissions that occur in a company's value chain, including purchased goods and services, transportation, business travel, and waste disposal.

For THE businesses, Scope 1 and 2 emissions are largely driven by on-site energy consumption (electricity, gas) and company vehicle use, while Scope 3 emissions are significantly influenced by purchased goods and services, particularly food procurement, and transportation (both guest and supply chain related).

Short-term measures (one to three years)

These measures focus on quick wins and behavioural adjustments with minimal capital investment.

Automated meter reading (AMR)

Implementing AMR systems allows for granular monitoring of energy consumption, often on a half-hourly basis. This detailed data facilitates the identification of energy waste hotspots (e.g. equipment left on overnight) and provides immediate feedback on the effectiveness of implemented changes. 'If it can be measured, it can be managed' is a key principle here.

Behavioural change programmes

Following energy audits and observations of daily practices, targeted behavioural change programmes can address specific areas of waste. A key strategy is to link desired behaviours with existing routines (e.g. adding an equipment check to the building closing procedure). Focusing on one behaviour at a time until it becomes ingrained is more effective than attempting to change everything at once.

Optimizing HVAC systems

Heating, ventilation and air conditioning (HVAC) systems, especially in larger establishments, are major energy consumers. Ensuring timers are correctly set and functioning, and that vents and grills are regularly cleaned and serviced, can significantly improve efficiency.

Transitioning to LED lighting

Replacing traditional lighting with LEDs offers a rapid and cost-effective way to reduce energy consumption and associated carbon emissions. The cumulative impact of widespread LED adoption across a property can be substantial.

Hot water management

Optimizing timers and temperatures on hot water boilers minimizes unnecessary heating and reduces energy waste. Addressing dripping taps and implementing regular boiler servicing further enhances efficiency.

Equipment timers

Installing timers on equipment like salamanders, heat lamps and toasters prevents them from being left on unnecessarily for extended periods.

Medium-term measures (three to five years)

These initiatives typically require more substantial investment and planning but offer greater potential for carbon footprint reduction.

Building retrofits

Incorporating energy-efficient equipment during renovations and refurbishments (e.g. induction hobs, improved insulation, double-paned windows) can significantly reduce long-term energy consumption. Retrofitting existing HVAC systems with Variable Air Volume (VAV) systems can also offer substantial energy savings.

Renewable energy procurement

Switching to renewable energy sources, either through direct purchase agreements or on-site generation, can drastically reduce Scope 1 and 2 emissions. However, it is crucial to remember that renewable energy procurement should complement, not replace, energy efficiency efforts. Reducing consumption helps balance demand with supply and prevents grid instability (International Energy Agency, 2023).

On-site renewable energy generation

Installing solar panels or other renewable energy technologies on-site can provide a long-term source of clean energy and potentially allow businesses to become net energy exporters.

Long-term measures (five to ten years and beyond)

These measures involve strategic planning and significant investment but are essential for achieving deep decarbonization.

Electrification of systems

Transitioning away from gas-powered equipment to electric alternatives requires careful assessment of existing electrical infrastructure and budgeting for equipment replacement. Consideration should also be given to the end-of-life management of replaced equipment.

Green building standards

Incorporating green building certifications and frameworks (e.g. LEED) into new construction projects ensures that buildings are designed and operated to minimize environmental impact.

Addressing Scope 3 emissions

Engaging with suppliers to reduce their emissions is crucial, particularly in the food procurement sector, where Scope 3 emissions can be dominant. This involves collaborative efforts to measure, reduce and report emissions throughout the supply chain.

Food procurement and menu formulation

Food procurement plays a critical role in Scope 3 emissions, especially in the THE industry:

- **Short term**: Introducing seasonal and local ingredients can subtly reduce the carbon footprint of menu items. Increasing plant-based options on menus also offers significant potential for emissions reduction.
- **Medium term**: Collaborating with suppliers to reduce their carbon emissions is essential. Understanding the carbon footprint of menu items allows for informed ingredient swaps. Carbon labelling can encourage customers to make more sustainable choices.
- **Long term**: Industry-wide collaboration is vital to drive systemic change in supplier practices and promote the adoption of sustainable agriculture. Investing in life cycle assessments (LCAs) for key ingredients can provide valuable insights for menu optimization.

Buildings and capital expenditure

Carbon-minimizing practices in construction and renovation offer both challenges and opportunities:

- **Short term**: Conducting insulation surveys, addressing draughts and promoting simple behavioural changes (e.g. closing doors when heating or cooling systems are in use) can improve energy efficiency.
- **Medium term**: Upgrading glazing in older buildings and prioritizing the reuse of existing fixtures and fittings during renovations can reduce embodied carbon and energy consumption.

- **Long term:** Exploring alternative construction materials with lower embodied carbon (e.g. sustainably sourced timber) and integrating on-site renewable energy generation and battery storage can minimize the carbon footprint of buildings.

Water reduction

Although water has a relatively low carbon impact compared to energy, it is a critical resource, and reducing consumption is an important part of broader ESG efforts:

- **Short term:** Addressing leaks and promoting water-saving behaviours among guests and staff (e.g. towel reuse programmes) can lead to immediate reductions in water usage.
- **Medium term:** Installing water-efficient fixtures (e.g. low-flow showerheads, dual-flush toilets) and PIR-activated urinals can significantly reduce water consumption.
- **Long term:** Implementing smart water management systems and exploring greywater recycling can further optimize water use and minimize environmental impact.

Transportation

Transportation is a significant contributor to the industry's carbon footprint, both for guests and supply chains:

- **Short term:** Promoting driver training programmes to improve fuel efficiency and encouraging the use of public transportation for longer journeys can reduce transportation-related emissions.
- **Medium term:** Engaging with suppliers to use electric vehicles or optimize delivery routes can minimize the environmental impact of supply chain transportation.
- **Long term:** Investing in electric vehicle fleets and providing charging infrastructure can significantly reduce transportation-related emissions.

Behavioural change: A critical component

Many of the aforementioned strategies require changes in behaviour, both for employees and guests.

Engaging customers

Nudging customers towards sustainable choices without compromising their experience is key. This can involve subtle cues and incentives.

Engaging teams

Creating a culture of sustainability within the workplace requires engaging all levels of the business. Clear and simple messaging, coupled with integrating new sustainable behaviours into existing routines, is essential. Regularly measuring and communicating the tangible impacts of these changes (e.g. cost savings, reduced emissions) to stakeholders reinforces the value of these efforts. For instance, rather than simply stating a site reduced Scope 1 and 2 carbon emissions by 2 USA tonnes (1.78 UK tonnes = 1,815 kilos), it should highlight practical outcomes, such as having saved 207 USA gallons (173 UK gallons = 784 litres) of fuel consumed, equivalent to the emissions from driving a car for just under 5,000 miles (which equals the emissions from driving from London to Ulanbaatar) (EPA, 2024).

Reducing carbon footprint is not merely an environmental imperative; it is also a business opportunity. By implementing the strategies outlined in this section, THE businesses can not only minimize their environmental impact but also enhance their brand reputation, attract environmentally conscious customers and improve operational efficiency. While the journey towards decarbonization requires commitment, investment and ongoing effort, the long-term benefits for both the planet and the bottom line are undeniable. By embracing a culture of continuous improvement and collaboration, the THE industry can play a leading role in the transition to a low-carbon economy (Wang, 2023).

The preceding sections have detailed actionable strategies for reducing a business's environmental impact. However, demonstrating the effectiveness of these efforts and building trust with stakeholders requires more than just internal action. Increasingly, consumers, investors and other stakeholders are demanding transparency and verifiable proof of sustainability performance. This is where the importance of international certifications comes into play.

International certifications: Validating sustainability efforts

Adopting international certifications such as EarthCheck, B Corp, LEED, ISO and Green Key is crucial for businesses in THE industries aiming to demonstrate their commitment to sustainability and gain credibility with stakeholders. In this section, we will explore the importance of these certifications, provide an overview of key certifications, and discuss the benefits and process of obtaining them. These certifications are formal processes by which a product, service or organization is assessed and verified as meeting specific sustainability criteria and standards. They provide a standardized and transparent framework for measuring and reporting on sustainability performance, enhancing trust and accountability (Sanderford, McCoy and Keefe 2018).

For instance, EarthCheck provides science-based certification for tourism businesses, while B Corp certification ensures companies meet high standards of social and environmental performance. LEED focuses on green building practices, ISO offers widely accepted process-based standards, and Green Key is a leading standard for excellence in the field of environmental responsibility and sustainable operation within the tourism industry. These certifications help businesses showcase their dedication to sustainable practices, which is increasingly important as travellers seek out environmentally responsible options. Certification not only enhances a company's brand reputation but also ensures compliance with global regulations and improves overall sustainability performance. By adhering to recognized standards, businesses can effectively communicate their environmental and social responsibility efforts, thereby building trust with customers, investors and the community.

The primary objectives of obtaining certification include enhancing brand reputation, ensuring regulatory compliance and continuously improving sustainability performance, which collectively contribute to a more sustainable and competitive tourism industry. These certifications act as a bridge between the internal actions discussed in the previous section and the external validation required to demonstrate genuine commitment to sustainability. They provide a clear and credible signal to stakeholders that the business is not only doing the work but also meeting recognized standards of best practice.

ISO 14001

The ISO 14001 standard for environmental management systems provides a framework for businesses to identify, manage and reduce their environmental impacts. It helps organizations improve resource efficiency, reduce waste, and lower costs by implementing systematic processes for environmental management. By adhering to ISO 14001, businesses can ensure compliance with environmental regulations, enhance their environmental performance and demonstrate their commitment to sustainability. This certification not only boosts a company's reputation but also fosters trust among stakeholders, including customers, investors, and regulatory bodies, by showcasing a proactive approach to environmental responsibility.

LEED certification

The Leadership in Energy and Environmental Design (LEED) certification is a globally recognized standard for green buildings. It guides businesses in designing, constructing and operating buildings that are environmentally responsible and resource efficient. LEED certification focuses on sustainable site development, water savings, energy efficiency, materials selection and indoor environmental quality. By achieving

LEED certification, businesses can reduce their environmental footprint, lower operating costs and create healthier workspaces. This certification not only enhances a company's sustainability credentials but also attracts environmentally conscious clients and investors, demonstrating a strong commitment to environmental stewardship.

Green Key

Green Key certification is an international eco-label awarded to tourism and hospitality establishments that commit to sustainable practices. It focuses on environmental management, waste reduction, energy and water conservation, and promoting eco-friendly activities. By achieving Green Key certification, businesses demonstrate their dedication to environmental responsibility, enhance their marketability and attract eco-conscious travellers.

B Corp

B Corp certification is granted to businesses that meet high standards of social and environmental performance, accountability and transparency. It evaluates a company's impact on its workers, customers, community and environment. B Corp certification helps businesses build trust with stakeholders, attract socially responsible investors and differentiate themselves in the market by showcasing their commitment to positive social and environmental change.

EarthCheck

EarthCheck is a globally recognized standard for sustainable tourism. It provides a science-based framework for benchmarking, performance measurement and independent auditing. By achieving EarthCheck certification, businesses demonstrate their commitment to environmental responsibility, enhance their marketability and attract eco-conscious travellers, (for more information on EarthCheck please refer to the real-world example on p160).

Benefits of international certifications

International certifications significantly enhance brand reputation by demonstrating a business's commitment to sustainability and responsible practices. They ensure regulatory compliance, helping businesses meet and exceed local and international environmental standards (UNFSS, 2022). Certifications like LEED, Green Key and B Corp differentiate businesses in a crowded market, making them more attractive to eco-conscious customers who prioritize sustainability in their purchasing decisions (Velaoras et al, 2025).

These certifications also drive continuous improvement in sustainability practices, fostering a culture of ongoing environmental responsibility and innovation (GRI, 2023). By achieving and maintaining these certifications, businesses build trust with stakeholders, including customers, investors and regulatory bodies, which can lead to a competitive advantage in the marketplace (UNFSS, 2022). Overall, international certifications are a powerful tool for businesses to showcase their dedication to sustainability, enhance their market position and contribute positively to the environment and society.

How to obtain international certifications

To obtain international certifications, businesses must follow a systematic approach (see Figure 5.5). The process begins with assessing current practices and planning necessary changes to align with certification standards (UNFSS, 2022). This involves conducting a thorough evaluation of existing operations and identifying areas for improvement. Once the assessment is complete, businesses implement the required changes, which may include adopting new technologies, enhancing resource efficiency, and improving waste management practices (EarthCheck, 2021). Documenting all processes and practices is essential, as it provides evidence of compliance with certification criteria (GRI, 2023).

Following implementation, businesses undergo third-party audits where independent auditors verify that the changes meet the required standards (UNFSS, 2022). Any non-conformities identified during the audit must be addressed promptly to achieve certification. Maintaining compliance requires regular reviews and updates

Figure 5.5 The process of obtaining international certifications

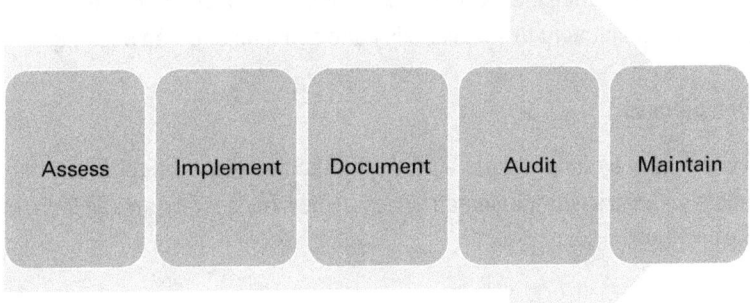

to ensure ongoing adherence to certification standards and continuous improvement in sustainability practices (EarthCheck, 2021).

These certifications act as a bridge between the internal actions discussed in the previous section and the external validation required to demonstrate genuine commitment to sustainability. Key concepts of this case study include data collection, benchmarking environmental data, sustainable reporting and independent audits, all crucial for demonstrating a genuine commitment to sustainability and avoiding greenwashing.

Having explored the importance and benefits of various international certifications, it is important now to explore a real-world example of how these principles are applied in practice by looking at the Athenaeum Hotel and Residences and their journey to achieving a global certification, in their case EarthCheck.

REAL-WORLD EXAMPLE EarthCheck and the Athenaeum Hotel & Residences

Kay Parfitt, Sustainability Manager at the Athenaeum Hotel & Residences, highlights the practical application of sustainability principles and the benefits of achieving international certification.

The Athenaeum Hotel & Residences exemplifies the practical application of sustainability principles and the tangible benefits of achieving international certification. By obtaining EarthCheck certification, the hotel has not only validated its sustainability efforts but also significantly enhanced its market position. This example underscores the importance of sustainability reporting and ethical practices in the hospitality industry.

Why EarthCheck?

EarthCheck was selected for its science-based approach and alignment with global standards such as IPCC and ISO. This certification provides a framework for benchmarking, performance measurement and independent auditing, which is crucial for demonstrating a commitment to sustainability and avoiding of greenwashing.

Certification process

The hotel underwent a comprehensive process to achieve EarthCheck Silver certification, including data submission, documentation across five key sectors and an independent audit to ensure transparency.

Outcomes and benefits

Since achieving certification, the Athenaeum Hotel & Residences has:

- reduced Scope 1 and 2 emissions by 73.81 per cent, demonstrating a significant environmental impact

- established an Environmental Management System (EMS), committing to continuous improvement
- gained recognition as a regional leader in waste management and carbon emission reduction

Key factors for success

The hotel's accomplishments can be attributed to:

- strong leadership commitment, driving the sustainability agenda from the top
- dedicated ESG management, ensuring focused and effective implementation of sustainability practices
- active engagement of staff through the 'Net-Zero Heroes' programme, fostering a culture of sustainability

Learnings and future directions

The hotel has faced challenges in data retrieval and team education, but these have been effectively addressed through tailored training and process improvements. Looking ahead, the focus is on measuring and reducing Scope 3 emissions to align with 2030 and 2050 sustainability targets.

Reflection

The Athenaeum Hotel & Residences example highlights sustainability as a journey requiring continuous effort, transparency and commitment. By partnering with reputable certification bodies like EarthCheck, businesses can validate their sustainability efforts and enhance their reputation.

> **REFLECTIVE QUESTIONS**
>
> 1. How did the Athenaeum Hotel & Residences ensure transparency and avoid greenwashing in their sustainability efforts?
> 2. What were the key steps in the hotel's journey to achieve EarthCheck Silver certification?
> 3. How did the hotel engage its staff in the sustainability initiatives?
> 4. What significant environmental impact did the hotel achieve post-certification?
> 5. What challenges did the hotel face, and how were they addressed?

ACTIVITY 1: SUSTAINABILITY AUDIT

This activity outlines the steps involved in a sustainability audit, from defining the scope and objectives to developing data collection methods and identifying potential improvements. It guides you through designing a practical plan, allowing you to apply your understanding of the key components and steps involved in conducting a sustainability audit.

This activity will take you through the process of designing a mini-sustainability audit. While you may not be able to implement it fully, the goal is to create a practical plan that could be put into action.

Instructions

1 **Define the scope and objectives**

   Choose a focus area: Select one area within a hotel or hospitality setting where you want to assess sustainability. Examples include:

   o guest room energy use
   o food waste in the kitchen
   o water usage in laundry
   o sustainable purchasing

   Set 2-3 clear objectives: What specific outcomes do you hope to achieve with your audit? (e.g. reduce energy consumption by 10 per cent, identify key sources of food waste)

2 **Develop data collection methods**

   Identify key metrics: What data will you need to collect to assess your chosen area? (e.g. energy bills, waste logs, water meter readings, supplier information)

   Create checklists or surveys: Design simple tools to gather your data. Example: Guest room energy checklist:

   o Are lights turned off when the room is empty?
   o Is the thermostat set to an energy-efficient temperature?
   o Are electronic devices unplugged when not in use?

   Plan observation methods: How will you observe and record relevant activities? (e.g. observe kitchen staff during food preparation to identify waste)

3 **Plan data analysis**

   How will you organize and analyze the data you collect? (e.g. calculate average energy use per room, track types and amounts of food waste)

What tools or methods will you use? (e.g. spreadsheets, basic statistical calculations)

4 **Identify potential improvements**

Based on your data and analysis, what specific actions could be taken to improve sustainability in your chosen area?

Example: If the audit reveals high energy use in guest rooms, potential improvements could include:

- installing occupancy sensors for lights and thermostats
- educating guests about energy-saving practices

Prioritize your recommendations based on feasibility and potential impact.

5 **Plan for follow-up**

How would you measure the success of your proposed improvements?

What steps would you take to ensure ongoing monitoring and continuous improvement?

Presentation

Prepare a brief presentation (5-10 minutes) outlining your mini-sustainability audit plan. Include:

- your chosen focus area and objectives
- data collection and analysis methods
- potential improvements and how you would measure their impact

The mini-sustainability audit activity is intended to consolidate students' understanding of sustainability measurement and reporting. By designing a practical audit plan, students reinforce theoretical knowledge of CSR, ESG and international certifications. They develop crucial skills in data collection, analysis, and interpretation, bridging the gap between academic learning and real-world application.

This task cultivates critical thinking, anticipatory, normative and self-awareness competencies, preparing students for effective decision-making in the industry. Simulating a real-world audit provides invaluable experience, ensuring students are well-equipped for future careers in THE.

Conclusion

This chapter has examined the importance of measuring impact and reporting in the tourism, hospitality and events (THE) industry. Key topics included ethical business practices, corporate social responsibility (CSR) and environmental, social and governance (ESG) frameworks. Practical tools like carbon footprint measurement and sustainability audits, and the role of international certifications in enhancing credibility, were also discussed. The Athenaeum Hotel & Residences real-world example illustrated the benefits of achieving certification, highlighting the importance of transparency, continuous improvement and stakeholder engagement. Finally, the students' activity on designing a mini-sustainability audit provided practical application of these concepts.

By integrating these principles, businesses can reduce their environmental impact, enhance their reputation and achieve long-term success.

KEY TAKEAWAYS

- Ethical business practices:
    - Ethical practices build trust, ensure legal compliance, enhance employee satisfaction and contribute to sustainable growth.
    - Key principles include environmental stewardship, respect for people, fairness, transparency and continuous improvement.
- Corporate social responsibility (CSR):
    - CSR involves contributing to sustainable development and improving the quality of life for stakeholders.
    - Key components include environmental sustainability, social responsibility, economic responsibility, ethical governance and stakeholder engagement.
- Environmental, social and governance (ESG):
    - ESG provides tangible metrics for assessing and managing sustainability issues.
    - It enhances transparency, accountability, and helps businesses manage risks and attract eco-conscious consumers.
- Carbon footprint and emission scopes:
    - Understanding Scope 1, 2 and 3 emissions is crucial for developing effective carbon reduction strategies.
    - Businesses can take targeted actions to reduce emissions and enhance sustainability performance.

- Sustainability audits:
  - Sustainability audits are essential for identifying inefficiencies, reducing resource use and improving profitability.
  - A structured approach to audits helps businesses achieve their ESG goals and demonstrate commitment to sustainability.
- International certifications:
  - Certifications like EarthCheck, B Corp, LEED, ISO and Green Key validate sustainability efforts and enhance brand reputation.
  - They ensure compliance with global standards and improve overall sustainability performance.
- Practical steps for implementation:
  - Implementing sustainability initiatives involves analysing current performance, setting clear goals and engaging stakeholders.
  - Continuous monitoring, reporting and improvement are essential for long-term success.

REFLECTIVE QUESTIONS

1 How do ethical business practices contribute to building trust and enhancing reputation in the THE industry?
2 In what ways can THE businesses effectively integrate CSR into their operations to benefit both the business and the community?
3 How does ESG reporting provide tangible metrics for transparency and accountability in THE businesses?
4 What are the key benefits and challenges for THE businesses in obtaining and maintaining international sustainability certifications?

Further resources

UKHospitality (2022) Environmental sustainability guide: For SMEs in the hospitality sector, https://app.sheepcrm.com/ukhospitality/member-only-documents/environmental-sustainability-guide/ (archived at https://perma.cc/QVV2-NVW7)

WRAP (.n.d.) HISTORY From Commitment to Transformation: The evolution of the UK Food and Drink Pact, www.wrap.ngo/take-action/uk-food-drink-pact/history (archived at https://perma.cc/7RQD-8M2V)

WSHA (2023) Net zero methodology for hotels, June, https://sustainablehospitalityalliance. org/resource/net-zero-methodology-for-hotels/ (archived at https://perma.cc/9MYE-8RCC)

WTTC (2017) Sustainability reporting in the travel & tourism sector, 1 September, https:// researchhub.wttc.org/product/environmental-social-governance-reporting-in-travel-and-tourism (archived at https://perma.cc/AX5V-DM5Q)

WTTC (2021) A net zero roadmap for travel & tourism, 20 November, https://wttc.org/ portals/0/documents/reports/2021/wttc_net_zero_roadmap.pdf (archived at https:// perma.cc/3SS9-9PT2)

References

Abul Basher Rasel, M (2024) Navigating sustainability in the tourism and hospitality industry: Transformation under the influence of climatic challenges, *SocioEconomic Challenges*, 8 (4), pp 112–20, https://doi.org/10.61093/sec.8(4).112-120.2024 (archived at https://perma.cc/V7MM-8XPQ)

Accor (2021) Accor Planet 21, www.accorhotelscomms.com (archived at https://perma.cc/ GXK4-EWNZ)

Anjum, W, Ghufran, B and Abbas, M (2024) Ecological footprint: A bibliometric analysis, *Environment, Development and Sustainability*, **2024**, pp 1–35, https://doi.org/10.1007/ s10668-024-05463-y (archived at https://perma.cc/6EDE-J6JV)

Ateeq, A and Milhem, M (2024) Integrating ethical principles in corporate strategy: A comprehensive analysis of business ethics in contemporary enterprises. In B Awwad (ed.) *The AI Revolution: Driving business innovation and research: Volume 1*, https://doi.org/ 10.1007/978-3-031-54379-1_17 (archived at https://perma.cc/ESQ3-LWQP)

Awa, H O, Etim, W and Ogbonda, E (2024) Stakeholders, stakeholder theory and Corporate Social Responsibility (CSR), *International Journal of Corporate Social Responsibility*, **9** (1), p 11, https://doi.org/10.1186/s40991-024-00094-y (archived at https://perma.cc/ UR6G-RYXU)

Bermudez, S, Vivek, V and Larrea, C (2022) Global market report: Coffee prices and sustainability, International Institute for Sustainable Development, September, www.iisd. org/system/files/2022-09/2022-global-market-report-coffee.pdf (archived at https://perma. cc/N5SH-EE8W)

Cambridge Institute for Sustainability Leadership (CISL) (2024) Competitive Sustainability Index: Shaping a new model of European competitiveness 'beyond Draghi', 10 December, www.cisl.cam.ac.uk/news-and-resources/publications/cisls-2024-competitive-sustainability-index-shaping-new-model (archived at https://perma.cc/6EN5-GEMD)

Climate Change Committee (2020) CCC Insights Briefing 3: The UK's Net Zero Target, October, www.theccc.org.uk/wp-content/uploads/2020/10/CCC-Insights-Briefing-3-The-UKs-Net-Zero-target.pdf (archived at https://perma.cc/MB4B-AU4E)

Coles, T, Fenclova, E and Dinan, C (2013) Tourism and Corporate Social Responsibility: A critical review and research agenda, *Tourism Management Perspectives*, **6**, pp 122–41, https://doi.org/10.1016/j.tmp.2013.02.001 (archived at https://perma.cc/JR84-79AG)

EarthCheck (2021) Tourism for inclusive growth, 27 September, https://earthcheck.org/news/ tourism-for-inclusive-growth/ (archived at https://perma.cc/UMH7-5JW2)

EPA (2024) Greenhouse Gas Equivalencies Calculator, www.epa.gov/energy/greenhouse-gas-equivalencies-calculator (archived at https://perma.cc/EMZ6-U2Y8)

Fatima, T and Elbanna, S (2022) Corporate Social Responsibility (CSR) implementation: A review and a research agenda towards an integrative framework, *Journal of Business Ethics*, 183 (1), pp 105–21, https://doi.org/10.1007/s10551-022-05047-8 (archived at https://perma.cc/7EV2-QBTD)

GHG Protocol (.n.d.) Standards & Guidance, https://ghgprotocol.org/standards-guidance (archived at https://perma.cc/E4EN-UDGT)

GRI (2023) GRI Annual Sustainability Report 2022: Towards a Global Comprehensive Reporting System, www.globalreporting.org/media/3yfhrjrk/gri-sustainabilityreport2022-final.pdf

Grüter, R, Trachsel, T, Laube, P and Jaisli, I (2022) Expected global suitability of coffee, cashew and avocado due to climate change, *PloS One*, 17 (1), e0261976, https://doi.org/10.1371/journal.pone.0261976 (archived at https://perma.cc/VT4C-7WWD)

Hasan, M B, Verma, R, Sharma, D, Moghalles, S A M and Hasan, S A S (2024) The impact of environmental, social, and governance (ESG) practices on customer behavior towards the brand in light of digital transformation: Perceptions of university students, *Cogent Business & Management*, 11 (1), p 2371063, https://doi.org/10.1080/23311975.2024.2371063 (archived at https://perma.cc/Y57Y-WAWY)

Hilton (2024) Environmental, Social and Governance at Hilton, https://esg.hilton.com/ (archived at https://perma.cc/Z6SN-F98J)

International Energy Agency (2020) Electricity Market Report 2023, OECD, 16 December, www.oecd.org/en/publications/electricity-market-report_f0aed4e6-en.html (archived at https://perma.cc/GRG4-A4X9)

Luque-Vílchez, M, Cordazzo, M, Rimmel, G and Tilt, C A (2023) Key aspects of sustainability reporting quality and the future of GRI, *Sustainability Accounting, Management and Policy Journal*, 14 (4), pp 637–59, https://doi.org/10.1108/SAMPJ-03-2023-0127 (archived at https://perma.cc/MH3G-GR2Y)

Marriott (.n.d.) SERVE360, https://serve360.marriott.com/ (archived at https://perma.cc/AV53-PJMC)

Molinos-Senante, M, Maziotis, A, Sala-Garrido, R and Mocholi-Arce, M (2022) Estimating performance and savings of water leakages and unplanned water supply interruptions in drinking water providers, *Resources, Conservation and Recycling*, 186, 106538, https://doi.org/10.1016/j.resconrec.2022.106538 (archived at https://perma.cc/2DQP-XUSP)

Net Zero Carbon Events (2022) A net zero roadmap for the events industry, 26 January, www.netzerocarbonevents.org/wp-content/uploads/NZCE_Roadmap2022_Full-Report-updated-26Jan2023.pdf (archived at https://perma.cc/6CES-PXSQ)

Ogbeide, G C A, Böser, S, Harrinton, R J and Ottenbacher, M C (2017) Complaint management in hospitality organizations: The role of empowerment and other service recovery attributes impacting loyalty and satisfaction, *Tourism and Hospitality Research*, 17 (2), pp 204–16, https://doi.org/10.1177/1467358415613409 (archived at https://perma.cc/5ZTD-H8ZW)

Operato, L, Gallo, A, Marino, E A E and Mattioli, D (2025) Navigating CSRD reporting: Turning compliance into sustainable development with science-based metrics, *Environmental Development*, 54, 101138, https://doi.org/10.1016/j.envdev.2025.101138 (archived at https://perma.cc/224D-CYAM)

Sanderford, A R, McCoy, A P and Keefe, M J (2018) Adoption of energy star certifications: Theory and evidence compared, *Building Research & Information*, **46** (2), pp 207–19, https://doi.org/10.1080/09613218.2016.1252618 (archived at https://perma.cc/877V-T422)

Steinmann, J (2024) Signs of a shift in business climate action, Deloitte, September 2024, www.deloitte.com/content/dam/assets-shared/docs/about/2024/deloitte-2024-cxo-sustainability-report.pdf (archived at https://perma.cc/Z3CM-AW5N)

TEAM (.n.d.) Understanding Scope 4 emissions and your carbon reduction plan, www.teamenergy.com/discover/blog/understanding-scope-4-emissions/ (archived at https://perma.cc/9WE8-PYGR)

Townsend, B (2020) From SRI to ESG: The origins of socially responsible and sustainable investing, *The Journal of Impact and ESG Investing*, **1** (1), pp 10–25, https://doi.org/10.3905/jesg.2020.1.1.010

UNDP (2024) Rethinking the governance of ESG, www.undp.org/future-development/signals-spotlight-2023/rethinking-governance-esg (archived at https://perma.cc/6RTD-XJVH)

UNESCO (2017) Education for Sustainable Development Goals Learning Objectives, https://unesdoc.unesco.org/ark:/48223/pf0000247444 (archived at https://perma.cc/DRU5-39HZ)

UNFSS (2022) Voluntary Sustainability Standards: Sustainability Agenda and Developing Countries: Opportunities and Challenges, https://unfss.org/wp-content/uploads/2022/10/UNFSS-5th-Report_14Oct2022_rev.pdf (archived at https://perma.cc/SNL4-2HLW)

UNWTO (1999) Global Code of Ethics for Tourism, www.unwto.org/global-code-of-ethics-for-tourism (archived at https://perma.cc/6F4Q-9D4Q)

UNWTO (2017) Manila Call for Action on Measuring Sustainable Tourism, 23 June, https://webunwto.s3-eu-west-1.amazonaws.com/imported_images/47298/call_for_action_filipinas.pdf (archived at https://perma.cc/RY44-X8ZP)

UNWTO (2024) Statistical Framework for Measuring the Sustainability of Tourism (SF-MST): Final draft prepared for UN Statistical Commission, www.unwto.org/tourism-statistics/statistical-framework-for-measuring-the-sustainability-of-tourism (archived at https://perma.cc/X6PE-36HT)

Velaoras, K, Menegaki, A N, Polyzos, S and Gotzamani, K (2025) The role of environmental certification in the hospitality industry: Assessing sustainability, consumer preferences, and the economic impact, *Sustainability*, **17** (2), p 650, https://doi.org/10.3390/su17020650 (archived at https://perma.cc/M2FA-GQTT)

Wang, H K H (2023) *The Roadmap for Sustainable Business and Net Zero Carbon Emission*, Routledge, London

WRAP (2024) Courtauld Commitment 2030 Annual Report 2023-2024, Autumn, www.wrap.ngo/sites/default/files/2024-12/241205-WRAP-Courtauld-2030-Annual-Progress-Report.pdf (archived at https://perma.cc/H4AE-5TLZ)

Zero Carbon Forum (2022) Net Zero: The guide for the brewing and hospitality sector, UK Hospitality, https://app.sheepcrm.com/ukhospitality/digital-docs/net-Zero-the-guide-for-the-brewing-and-hospitality-sector/ (archived at https://perma.cc/2H6G-D6FT)

Zhang, K and Hao, X (2024) Corporate social responsibility as the pathway towards sustainability: A state-of-the-art review in Asia economics, *Discover Sustainability*, **5** (1), p 348, https://doi.org/10.1007/s43621-024-00577-9 (archived at https://perma.cc/9EV4-UM3S)

6 | The circular economy

| Systems thinking competency | Critical thinking competency | Strategic competency | Collaboration competency |

CHAPTER AIM

This chapter aims to develop a critical and applied understanding of the circular economy and its importance in addressing climate change within the tourism, hospitality and events (THE) industries. It introduces the theoretical foundations of circular economic models and explores how these principles can be implemented to drive sustainable transformation across THE sectors. Emphasis is placed on aligning circular practices with key global frameworks, including the SDGs – specifically SDG 11, SDG 12, SDG 13 and SDG 17.

The chapter supports the development of key sustainability competencies as identified by UNESCO (2017): systems thinking, critical thinking, strategic and collaboration competencies. Through real-world examples, project-based learning and reflective practice, students will explore how circular economy strategies can be embedded into THE operations, destination planning and experience design. The chapter encourages learners to think holistically, question prevailing practices and identify innovative, cross-sectoral solutions for a more sustainable and resilient future in THE.

LEARNING OUTCOMES

Upon completion of this chapter, you will be able to:

- define and explain the core principles of the circular economy in the context of tourism, hospitality and events
- critically assess the limitations of linear models and identify opportunities for circular innovation within THE sectors
- apply systems thinking to analyze interdependencies between economic, environmental, and social systems in THE
- design strategic responses to sustainability challenges using circular economy approaches
- demonstrate collaborative problem-solving skills by evaluating stakeholder roles and cross-sector partnerships essential to circular transitions
- reflect on personal and professional practice in relation to circular economy principles and sustainability principles

KEY WORDS

Circular economy, net zero, linear economy, closed-loop system, 8Rs framework, waste reduction, cross-industry collaboration.

Introduction

The climate emergency represents one of the most pressing challenges of our time. Climate change, driven by greenhouse gas (GHG) emissions, has caused rising global temperatures, extreme weather patterns and biodiversity loss (IPCC, 2023).

The Intergovernmental Panel on Climate Change (IPCC) warns that global temperatures have already increased by 1.1oC above pre-industrial levels, with projections indicating severe consequences if warming surpasses 1.5oC (IPCC, 2023). The THE sector, which contributes approximately 8 per cent of global GHG emissions (Lenzen et al, 2018) plays a significant role in this crisis as Figure 6.1. shows. THE is heavily reliant on natural and cultural resources, energy use and waste generation, with high carbon emissions from transport and operations. Aviation alone accounts for 2.5 per cent of global CO_2 emissions, with additional emissions generated by cruise ships, hotels and infrastructure development (International Air Transport Association, .n.d.). The carbon footprint of tourism is not only comprised of transport, accommodation, and goods and beverage, but also includes the activities of tourists including buying souvenirs and embarking on excursions.

Addressing these issues requires THE to act by developing net zero strategies and circular economy principles, to mitigate climate impacts and ensure long-term sustainability. It is useful to be reminded of the meaning of the key terms being used in this chapter.

Greenhouse gases (GHGs) and net zero

Greenhouse gases (GHGs) are atmospheric gases that trap heat within Earth's atmosphere. These gases allow sunlight to pass through, but prevent the heat generated by sunlight from escaping back into space. The primary greenhouse gases include water vapour, carbon dioxide, methane, ozone, nitrous oxide and chlorofluorocarbons (CFCs). They derive their name from their similarity to a greenhouse, where sunlight enters through glass windows to create warmth for plant growth, but the structure prevents heat from escaping.

This heat-trapping effect is essential for maintaining Earth's habitable climate. Without greenhouse gases, the planet would be too cold to sustain life. However, human activities such as burning fossil fuels, deforestation and industrial processes have significantly increased the concentration of GHGs in the atmosphere. This excess accumulation is causing global concerns due to its impact on climate change, leading to rising global temperatures and disrupted ecosystems (NASA, .n.d.). GHG emissions, primarily from fossil fuels such as coal, oil and gas, contribute to approximately 68 per cent of total global GHG emissions, with THE related industries such as transport and agriculture contributing an additional 26 per cent (UNEP, 2024). The United Nations World Tourism Organization (UNWTO) launched the Glasgow Declaration on Climate Action in Tourism in 2021, urging business and destinations to cut emissions by 50 per cent by 2030, and achieve net zero (see below) by 2050 (UNWTO, 2021). The United Nations Environment Programme (UNEP) Emissions Gap Report 2024 reiterates UNWTO's call by pointing out that

Figure 6.1   The carbon footprint of global tourism

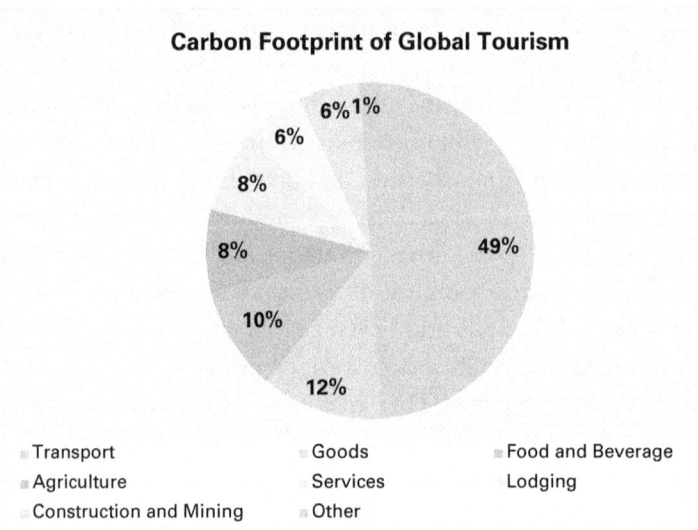

The percentage of different activities that contributed to the global tourism carbon footprint in 2018. Data from Lenzen et al, 2018

reductions in emissions of 42 per cent are needed by 2023 and 57 per cent by 2035 if the global target for no more than 1.5oC increase in temperature is to be met.

Net zero emissions refers to balancing emitted GHGs with their removal from the atmosphere, a critical step in limiting climate change. Net zero means that a company, community or country is not adding any more greenhouse gases to the atmosphere than it is taking away. This can be achieved through a combination of reducing emissions and enhancing carbon sinks, such as forests and oceans, which absorb more CO_2 from the atmosphere than they release, helping to mitigate climate change (Broom, 2023).

To address these challenges and meet the net zero target requires a shift in how THE operates. As an industry THE needs innovative solutions, particularly in resource consumption and waste production, to reduce greenhouse gases effectively.

By integrating circular economy principles, systems can be created that not only reduce THE's carbon footprint but also support the regeneration of natural resources, helping achieve net zero goals.

Integrating circular economy principles in THE

Traditionally THE has adopted a linear market economy approach to its business operations. In this system, raw materials are extracted, transformed into products, used by consumers, and then discarded as waste, with little or no emphasis on recycling or reuse. This linear progression – from resource extraction to disposal – is focused on minimizing costs and maximizing short-term profits, often at the expense of sustainability (Knight, 2023).

The circular economy offers an alternative to the traditional linear economy by focusing on minimizing waste and maximizing resource efficiency through the principles of reduce, reuse, recycle. This model aims to conserve natural recourses, reduce pollution and prevent the over-extraction of raw materials by incorporating secondary raw materials – products designed to be repaired, upgraded or remanufactured at the end of their life cycle. Closely linked to the circular economy is the sharing economy, which emphasizes collaboration over ownership. In THE this can involve initiatives such as shared transport, accommodations and services, to enable more sustainable and resource-conscious practices across the industry.

To understand the relevance of the circular economy for THE, it is essential to first examine the globally dominant linear economy model.

The linear economy

The linear economy is the dominant economic model based on a 'take, make, dispose' approach (see Figure 6.2). Raw materials are extracted, transformed into products, used, and then discarded. This model emerged during the Industrial Revolution, driven by an abundance of natural resources and the emphasis on mass production. The linear economy thrived under the assumption of limitless resources, leading to the widespread environmental and resource depletion issues experienced in contemporary economies.

This process is straightforward but unsustainable. For example, in the case of an aluminium drink can, the metal is mined, processed into a can, used, and then disposed of, often ending up in landfill or incineration. This cycle results in significant resource wastage and environmental impact.

The linear economy relies on the assumption of unlimited resources and has limited incentives for recycling, reuse or reducing waste (see Figure 6.3). As a result, the linear model contributes to resource depletion and environmental degradation, posing long-term sustainability challenges for THE (Manniche et al, 2018).

Figure 6.2 The linear economy cycle

TAKE → MAKE → WASTE

Figure 6.3 Linear economy cycle assumption

Material extraction → Manufacture → Distribution → Usage → End of life

Recognizing the limitations of the linear economy highlights the need for alternative models. The circular economy offers a more sustainable and resource-efficient approach, addressing the challenges of waste generation and resource depletion through a closed-loop system.

A closed loop system in THE

A closed-loop system in contemporary terms refers to a production and consumption model in which resources are continuously reused, remanufactured, recycled or composted, thereby minimizing waste and reducing the extraction of raw materials. In such systems, the output (waste or byproducts) is fed back into the system as input, forming a regenerative cycle (Ellen MacArthur Foundation, 2021).

In the context of THE, a closed loop system refers to a sustainable operational model in which materials, energy and resources are continuously recycled within the system of service delivery. It involves designing tourism activities so that waste products such as food waste, packaging, energy or construction materials are further reused, recycled or returned to nature through composting (Amicarelli et al, 2022). A closed-loop system is a practical strategy through which THE businesses can apply circular economy principles in their daily operations.

The circular economy

The circular economy provides a framework for transitioning from a linear model of consumption – characterized by a 'take, make, dispose' approach – to one that emphasizes resource efficiency, waste reduction and material reuse. Natural cycles in nature, such as the water, carbon and nutrient cycles, demonstrate the effectiveness of closed-loop systems in maintaining ecological balance. The circular economy mirrors these natural cycles by ensuring that resources are continuously repurposed, reducing dependency on raw material extraction and minimizing environmental impact. Breaking these cycles disrupts efficiency, much like a bicycle with a buckled wheel – functional but significantly impaired. A complete break halts the system entirely, as seen in the inefficiencies of the current linear economic model.

By minimizing waste and extending the lifespan of materials, the circular economy reduces GHG emissions and environmental impact (Manniche et al, 2018). The circular economy also aims to maximize the use of manufactured products by sharing and leasing instead of owning them: turning products into services. In THE, adopting strategies such as regenerative tourism, waste reduction and resource efficiency allows businesses and destinations to contribute to environmental conservation while building sustainable economic models (Manniche et al, 2018).

Figure 6.4 The 8Rs framework

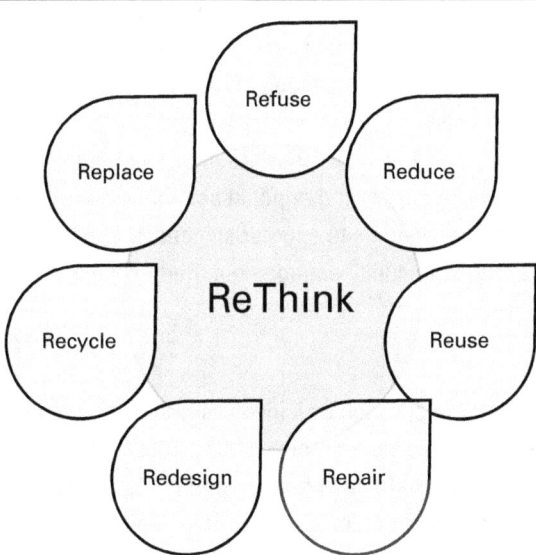

A fundamental principle of the circular economy is the application of waste reduction strategies. While the 3Rs – reduce, reuse, recycle – form the foundation of sustainability, additional strategies enhance resource efficiency within this model. The 8Rs framework provides a structured approach to circularity: placing 'rethink' at the centre of the framework with the remaining behaviors organized around 'rethinking': refuse, reduce, reuse, repair, redesign, recycle and replace (see Figure 6.4). These steps guide sustainable decision-making and business practices, ensuring resources remain in use for as long as possible, extracting maximum value before recovery and regeneration (Ellen MacArthur Foundation, 2021).

To understand the 8Rs framework, an explanation of each R is supported by an example from one hotel – the Hotel Casa Palmela in Portugal.

REAL-WORLD EXAMPLE Exploring the 8Rs framework: The Hotel Casa Palmela in Arrábida Natural Park, Portugal

The following example of the integration of circular economy principles in a hotel provides a real-world example of a sustainable model that enhances both environmental stewardship and operational efficiency. Through the adoption of the 8Rs framework, the luxury nature hotel Hotel Casa Palmela in Arrábida Natural Park, Portugal, exemplifies how sustainable practices can be implemented across various aspects of hospitality operations.

Rethinking

Rethinking requires THE businesses to reflect upon what they do, where they do it and why they need to do it. This involves reassessing current consumption and production patterns to integrate circularity at every stage. It requires consideration of how products are designed, used and repurposed to maximize longevity and minimize waste. In THE, hotels can rethink their supply chains to prioritize local, sustainable products, reducing transportation emissions and supporting regional economies. Tour operators can shift from traditional mass tourism models to ecotourism, ensuring a lower environmental impact while engendering community engagement. Rethinking is present in every phase of the circular economy.

Redesigning

This is the most difficult R to approach as it involves rethinking product and service design to prioritize sustainability, durability and recyclability. This step requires innovation to create solutions that reduce waste and enhance efficiency. Event venues can be designed to be modular and adaptable, allowing for multiple configurations that reduce renovation waste.

However, redesigning can present challenges for historical buildings that may be listed and have restrictions placed upon their preservation. While the structure of an old hotel, for example, may be difficult to redesign, the interiors can use locally sourced and recycled materials and eco-friendly paints and upcycle fixtures and fittings. Redesigning products and services for sustainability also requires investment, but can lead to long-term cost savings and environmental benefits.

The Hotel Casa Palmela has rethought its traditional hospitality practices by embedding sustainability into its operations. This involved not only minimizing negative environmental impacts but also designing systems and services that contribute positively to the environment and society. For instance, the hotel redesigned its infrastructure to incorporate energy-efficient lighting, water-saving measures and environmentally friendly building materials during its renovation. Thermodynamic solar panels for hot water, low-flow taps and showers, LED lighting and motion sensors have all been integrated to reduce energy and water consumption. By adopting these innovative solutions, the hotel has reduced its reliance on raw materials and promotes long-term resource efficiency, aligning with the circular economy's emphasis on sustainability through rethinking for redesigning.

Refusing

Avoiding unnecessary consumption is key to reducing waste. Businesses and consumers can refuse single-use items, opting for durable alternatives. For instance, event organizers can eliminate disposable cutlery in favour of reusable options or restructure food offerings to minimize packaging waste. Airlines and hotels can refuse single-use plastic amenities, replacing them with sustainable alternatives such as biodegradable packaging or refillable dispensers.

However, the goal would be not to make the items in the first place, or at least make less of them. The circular economy is driven to make less and what is made should be used in its highest form for as long as possible. Fewer resources used means lower energy consumption, lower energy is likely to mean less fossil fuel being burned, and this is turn leads to less carbon dioxide. For example, to refuse use of cutlery altogether, event organizers could opt for food options such as 'finger food' that require no cutlery.

The principle of refusal is reflected in Hotel Casa Palmela's practices aimed at minimizing overconsumption and waste. One key action has been the reduction of single-use plastics and non-recyclable materials by opting for reusable articles. For example, the hotel reuse their water bottles on a daily basis by washing them and filtering their own water to refill them (Casa Palmela Hotel, .n.d.) This proactive approach reduces waste and helps minimize the hotel's ecological footprint, supporting a circular economy model where resource consumption is kept to a minimum.

Reducing

Consumption and waste generation must be minimized. Reducing material usage not only decreases environmental impact but also enhances efficiency. A key example is food waste reduction. According to UNEP's Food Waste Index Report 2024 (UNEP, 2024), the annual global food waste figure for 2022 was 1.05 billion tonnes, equating to one billion meals per day being wasted. THE significantly contributes to this issue, making it crucial to adopt strategies aligned with SDG 12, which targets halving food waste by 2030.

A rethink is necessary to achieve a balance, considering geographical, cultural, social, financial and environmental factors. Restaurants can reduce food waste by improving inventory management, utilizing entire fruits and vegetables rather than discarding stems or skins, and offering smaller portion sizes. Hotels can implement water-saving devices to reduce consumption, while airlines can optimize flight routes and adopt lighter aircraft materials to decrease fuel usage. Additionally, sourcing local products helps reduce transport emissions, and minimizing packaging lowers plastic waste.

Returning to the Hotel Casa Palmela as an example, it works to reduce its energy and water consumption by adopting efficient technologies. The use of low-flow fixtures, energy-efficient flushing equipment and thermodynamic solar panels for hot water all contribute to significant reductions in resource use. For example, thermodynamic solar panels can generate hot water up to 55°C (Pereira, Silva and Dias, 2021). This focus on reducing resource inputs is in line with circular economy principles, which advocate for minimizing waste and lowering environmental impact through smarter resource management.

Reusing

Reusing is a key principle of the circular economy that focuses on extending the life cycle of products and materials by using them multiple times instead of discarding them after a single use. This reduces waste, conserves resources and minimizes the environmental impact by keeping the products and materials in circulation for as long as possible.

In the hospitality industry for example, hotels can implement reuse practices by replacing single-use plastics with refillable dispensers for toiletries. This not only reduces plastic waste and operational costs, but also aligns with changing guest expectations and behaviours. Hotels can reuse linens and towels by encouraging guests to participate in linen reuse programmes which reduce the frequency of laundering, saving water and energy. It also encourages guests to rethink their behaviours. Attractions and resorts can establish deposit-return schemes for reusable cups and packaging to minimize waste generation.

Repairing

Repairing broken items is a fundamental aspect of the circular economy, extending the life cycle of products and minimizing waste. Instead of discarding damaged goods, seeking expert repairs can conserve valuable resources and reduce the environmental impact associated with manufacturing new items. Designers and manufacturers also play a crucial role in repairability by ensuring that products are built to last, with replaceable components and materials that can be repurposed. Encouraging repair over replacement aligns with sustainable business practices, reduces carbon footprints and helps preserve finite resources.

Within THE, repair strategies can be applied in various ways. Hotels and resorts can prioritize repairing furniture, fixtures and appliances instead of replacing them, reducing resource consumption and operational costs. Maintenance teams can be trained to fix common issues, ensuring equipment longevity. Additionally, outdoor tourism businesses such as adventure tour operators can implement repair programmes for gear and equipment like tents, bicycles and kayaks, reducing waste and promoting sustainability.

Destinations can further support these initiatives by supporting repair networks and workshops, ensuring tourists and businesses have access to affordable and skilled repair services. Additionally, prioritizing repair work creates employment opportunities in local communities by supporting tradespeople and technicians specializing in maintenance and refurbishment. This strengthens local economies while reinforcing sustainability efforts. Developing repair initiatives within THE can also promote traditional artisanry, ensuring valuable skills are preserved and adapted for modern applications.

Recycling

Recycling is the process of recovering and repurposing materials at the end of their life cycle to create new products. While recycling is essential, it should be considered a last resort after all other Rs have been exhausted. It is a difficult process because its success is influenced by geography, local infrastructure and services, communication, producer responsibility, and finances.

For example, looking at national parks as tourist attractions, the geography could relate to where recycling stations can be set up throughout the park with the purpose of encouraging visitors to sort their waste. This requires clear and precise communication including ensuring labelling on each recycling bin specifies recyclable categories like paper, plastic and aluminium. THE would need to form partnerships with recycling operatives to ensure that discarded materials are properly processed. All these considerations have financial implications; therefore, they would need to be included in an operational budget.

Hotel Casa Palmela's transition from a private estate to a hotel demonstrates the implementation of a reuse, repair, recycle approach. Sustainable construction practices, the repair and reuse of materials and the use of eco-friendly materials were used in the redevelopment of the hotel. Specifically, the hotel reused furniture and spaces to avoid a negative climate impact on the surrounding environment when undergoing its refurbishment (Casa Palmela Hotel, .n.d.).

A comprehensive waste management system plays a critical role in the hotel's circular economy strategy. By sorting and recycling materials like paper glass, and plastics, the hotel reduces the volume of waste sent to landfills. This practice supports a closed-loop system where materials are kept in circulation and reused, minimizing the need for virgin raw materials and reducing overall environmental impact.

In its daily operations, the restaurant's menus source from the hotel's own produce; they are designed around the seasons. For example, the summer months' menu is cold food-based. Food left over from breakfast is used again for the lunch menu. Food scraps are composted to produce fertilizer for the hotel's gardens and allotment. Damaged or broken articles used for decoration and service are also repaired or recycled by local expert potters (Casa Palmela Hotel, .n.d.).

Replacing

Replacing intersects with several of the other key Rs as it serves as an opportunity to pause and reconsider actions. It focuses on substituting resource-intensive products with sustainable alternatives. This often involves moving away from materials that are difficult to recycle or dispose of. For example, tour operators can rethink their use of traditional printed brochures by replacing them with digital versions. This not only reduces paper waste and avoids the use of contaminating inks, but also allows for more efficient updates. Restaurants can replace single-use condiment sachets with refillable containers. However, replacing unsustainable products with greener alternatives should be done gradually and strategically to minimize waste and ensure long-term sustainability. A well-planned transition prevents unnecessary disposal of existing materials and allows businesses to adhere to any local regulatory compliance.

The example of 'replacing' illustrated by the Hotel Casa Palmela is in the form of excursions. The hotel's guests are encouraged to engage with experiences offered on the estate itself or in conjunction with the local community. This practice avoids travel and reduces the environmental impact associated with fossil fuels for longer trips (Casa Palmela Hotel, .n.d.).

The 8Rs framework provides a comprehensive approach to embedding circular economy principles within THE. Rethinking, at the centre of the framework, is the driver for the remaining Rs because it not only requires thought but behavioural change. The practicality of implementing a circular economy emphasizes changing business and consumer habits; these changes are not always straightforward. However, the examination of the Hotel Casa Palmela's mission and management practice is indicative of the steps that can be taken.

The Hotel Casa Palmela's commitment to sustainability is exemplified by its replacement of traditional resource-intensive practices with more sustainable alternatives. In its daily resource management practice, it has instigated measures to reduce its emissions. This is being implemented through its non-use of fossil fuels and choice to use cleaner, renewable sources of electricity. Its circular economy principles are typified by:

- rethinking its traditional hospitality practices
- redesigning its accommodations and spaces
- refusing single-use packaging and items
- reducing energy and water consumption
- reusing refillable water bottles and bathroom toiletry containers
- repairing articles in daily use for service and decoration
- replacing excursions with local experiences
- recycling organic waste, glass, paper, metal and plastic

The Hotel Casa Palmela offers valuable insights for hotel owners and managers seeking to integrate sustainability into their operations. Through ecological conservation and sustainable practices, the hotel has successfully regenerated its local ecosystem and reduced its environmental footprint to contribute to long-term sustainability. It reports a 40 per cent reduction in energy and water consumption compared to its 2019 budget forecast, the minimization of its carbon footprint through reduced emissions, and a smaller ecological impact in line with the conservation goals of the surrounding Arrábida Natural Park. The hotel's commitment to social sustainability has also yielded positive outcomes, including enhanced community engagement and improved employee welfare. These efforts have advanced the hotel's reputation and attracted environmentally conscious guests to establish a competitive edge in the luxury hospitality market (Pereira, Silva and Dias, 2021).

The Hotel Casa Palmela exemplifies the transformative potential of adopting the 8Rs framework of the circular economy. By aligning its operations to the 8Rs, the Hotel Casa Palmela is an example of the regenerative nature of the circular economy. It also demonstrates how adopting a circular economy approach benefits not only the environment, but also the bottom line.

> **REFLECTIVE QUESTIONS**
>
> 1 What is the primary goal of Hotel Casa Palmela's circular economy initiatives?
> 2 How does Hotel Casa Palmela support local economies through its sustainability initiatives?
> 3 What strategies has Hotel Casa Palmela implemented to reduce food waste?
> 4 What are some of the energy efficiency measures taken by Hotel Casa Palmela?
> 5 How does guest engagement contribute to Hotel Casa Palmela's sustainability efforts?

While this example has focused on a hotel, it is evident that the adoption of circular economy principles can lead any THE business to significant operational efficiency and cost savings. The hotel demonstrates that sustainability practices are an investment rather than an additional expense, as they offer measurable long-term financial returns. In this regard, it is a model for other businesses in the THE sector.

These efforts have advanced the hotel's reputation, attracted environmentally conscious guests and established a competitive edge in the luxury hospitality market. Other leaders in the hotel industry echo this perspective, showing that glamour and environmental responsibility can not only coexist, but reinforce each other.

> **INDUSTRY VOICE: SALLY BECK**
>
> Sally Beck, general manager at Royal Lancaster London, says 'Sustainability doesn't mean giving up glamour. Quite the opposite – we want to show guests that ethical choices can enhance their experience. At the Lancaster, we're constantly looking at how to weave environmental responsibility into elegance. It's about showing that style and substance are not mutually exclusive.'

Sustainability in the circular economy

The transition to a circular economy presents an opportunity for THE. In the THE industries, where large scale operations often generate significant waste and emissions, sustainable production plays a central role as operations shift their focus from the linear model of consumption to a regenerative system. By integrating circular economy principles with net-zero strategies, THE industries can reduce environmental harm without undermining business operations. In practice this means focusing on sustainable production practices that minimize environmental impact at every stage of the product life cycle and support resource conservation. The emphasis is placed upon renewable materials, energy efficiency and waste reduction through the production life cycle.

To understand how sustainability is embedded within a circular economy, it is necessary to examine the key processes that shape this closed-loop system.

Resource extraction

In traditional systems, resource extraction has been conducted in ways that result in significant environmental degradation, including deforestation, pollution and depletion of non-renewable resources. Under the circular economy, resource extraction is minimized by ensuring that materials already in circulation are reused or recycled (Manniche et al, 2018). Through this process, the use of raw materials is reduced, and the environmental damage caused by mining and other extraction activities is lessened.

Progress in industries such as plastics has demonstrated how recycling efforts can reduce the demand for raw materials while also cutting energy consumption and carbon emissions. In the THE industry, wasteful practices are being addressed by replacing single-use plastics, such as toiletries, with refillable dispensers. By adopting such changes, environmental impacts are reduced and the move toward sustainability is supported.

The Circular Economy (Scotland) Act 2024 reinforces this direction by introducing provisions for ministers to set targets for material consumption and environmental impact reduction. The Act includes statutory measures to drive the reuse and recycling of products and encourages public and private sectors to adopt sustainable procurement practices. Notably, it places legal responsibility on producers and retailers to contribute to Scotland's circular ambitions, directly impacting sectors such as hospitality and tourism that rely on large-scale purchasing and resource use.

The negative impacts of resource extraction, including habitat destruction, water pollution and displacement of communities, are also mitigated when this approach is adopted. Policies such as the UK's Extended Producer Responsibility (EPR) legislation have been introduced to ensure businesses take responsibility for the full life cycle of their packaging, encouraging more sustainable practices (Department for Environment, Food & Rural Affairs, 2022).

Waste reduction and emissions

A circular approach means minimizing waste that would otherwise be directed to landfills, where decomposition processes release methane, a greenhouse gas with a significantly higher global warming potential than carbon dioxide (Environmental Literacy Council, 2024). Furthermore, waste incineration contributes to carbon dioxide emissions and the release of additional pollutants that adversely impact air quality (Unegg et al, 2023). The implementation of circular economy strategies, including composting, recycling and reducing waste at the source, is essential in lowering methane emissions and mitigating the negative environmental consequences associated with conventional waste disposal practices.

A key framework for effective waste management is the waste hierarchy model, as illustrated in Figure 6.5. This model, which is based on DEFRA guidance, is particularly relevant to THE as it prioritizes waste prevention, followed by preparation for reuse, recycling, recovery, and, as a last resort, disposal. In this regard, the model reflects the 8Rs framework. The adoption of this structured approach can substantially reduce waste generation, limit methane emissions from organic waste and mitigate the harmful effects of landfill disposal. For example, strategies such as minimizing single-use plastics, repurposing materials and enhancing recycling efforts yield significant environmental and economic benefits.

The Circular Economy (Scotland) Act 2024 supports this approach by giving new powers to local authorities to introduce a charge for the collection of household waste and to restrict the disposal of unsorted waste, incentivizing proper recycling and reducing contamination in waste streams. For THE businesses, especially in urban areas or destination hotspots, such provisions reinforce the importance of pre-sorting waste, investing in staff training, and implementing waste-reduction policies that align with both environmental and legal obligations.

The Waste (England and Wales) Regulations 2011 establish a legal framework to ensure businesses, including those operating within the THE industry, comply with waste hierarchy principles to minimize environmental impact (legislation.gov.uk, 2011). Adherence to these regulations is critical for organizations engaged in waste management, as it leads to sustainable operational practices and aligns with broader environmental objectives.

An illustrative example of effective waste reduction and emissions mitigation can be observed in the implementation of mobile applications designed to reduce food waste. Such applications facilitate the redistribution of surplus food from businesses to consumers by offering discounted 'surprise bags' containing unsold food items (Too Good To Go ApS, 2024). This initiative not only reduces food waste on a substantial scale but also raises awareness regarding sustainable food consumption. Moreover, by quantifying the environmental impact and promoting engagement from consumers, businesses, and regulatory bodies, such applications contribute to a broader social movement aimed at responsible resource utilization. While concerns regarding food safety, service disruptions and consumer trust must be addressed, these platforms play a significant role in reducing waste and mitigating food insecurity.

Figure 6.5 A waste hierarchy model

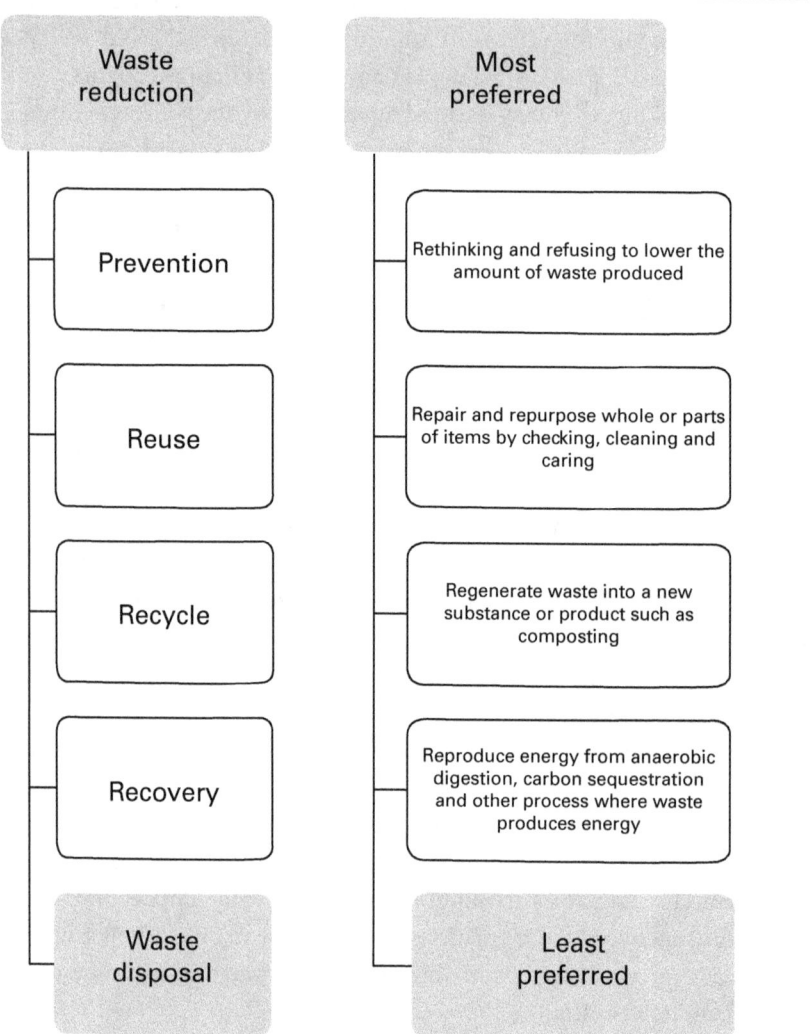

By systematically integrating sustainable practices from resource extraction to waste management – and aligning with legislative frameworks such as the Circular Economy (Scotland) Act 2024 – businesses can contribute to the development of a closed-loop system that minimizes environmental degradation.

Lowering energy consumption

Energy-intensive production processes have traditionally contributed significantly to greenhouse gas emissions. In the circular economy, lower energy consumption is achieved by improving energy efficiency and adopting renewable energy sources.

While the THE industry does not directly contribute to large-scale manufacturing, indirect contributions are made through the selection of products. For instance, products made from recycled materials require less energy to produce compared to those made from new raw materials. It has been found that producing recycled plastic uses 25–30 per cent less energy than creating new plastic from fossil fuels (Association of Plastic Recyclers, 2018). By selecting such materials, THE businesses contribute to energy savings and reduce environmental harm.

This reduction in energy consumption has also been linked to decreased greenhouse gas emissions and lower dependence on fossil fuels, aligning with sustainability goals and economic resilience.

Sustainable purchasing decisions

Purchasing decisions play a critical role in supporting sustainability efforts. The starting point for any purchase is to assess the necessity of making the purchase. In a circular economy, purchases should be made only when necessary to avoid unnecessary resource use. In addition to understanding the real necessity behind a purchasing decision, other circular economy principles should be applied including ethical considerations of the wider social impact of operations along the supply chain:

- **Life cycle assessment:** The circular economy encourages organizations to conduct life cycle assessments of products to understand their total carbon footprint. This understanding enables better decision-making regarding material choices, production processes and end-of-life options. For example, when investing in table linen eateries need to assess washing practices that will have a low energy consumption, such as being able to launder at cooler temperatures.
- **Quality and durability:** Linked to life cycle, products should be chosen for their longevity. Longer-lasting items require fewer replacements, reducing waste over time. Quality and durability can support tourism as they are valued by customers and contribute to their long-term satisfaction (Baloch et al, 2022).
- **Environmental impact:** Items made from eco-friendly materials or produced using energy-efficient methods should be prioritized. Certifications such as Forest Stewardship Council (FSC) or Energy Star can assist in identifying sustainable options (Baloch et al, 2022).
- **Ethical practices:** Considerations of social responsibility should also inform purchasing decisions. Items produced by companies with fair labour practices and responsible sourcing policies should be supported (Baloch et al, 2022). The Fairtrade Foundation, for example, sets social as well as economic and environmental standards for companies and farmers involved in the supply chain. Through FLOCERT, an independent organization, the farmers, workers and companies in

the supply chain receive a licence if fair trade social standards are met in conjunction with standards for ensuring environmentally sustainable practices and delivering economic benefits to local communities (Fairtrade, .n.d.).

Extending product life cycles – uniforms

Uniforms play a significant role in THE, both operationally and culturally. As part of the industry's shift toward circular economic principles, there is increasing attention on how uniforms are produced, used and disposed of. The aim is to reduce environmental impact, conserve resources and extend the useful life of products. Each year approximately 920,000 tonnes of textile waste is sent to landfill. A significant portion of this waste comes from synthetic materials like polyester, which are slow to degrade and release harmful microplastics into the environment. Given the high turnover and branding needs associated with uniforms in THE, the sector contributes directly to this issue. A circular approach to uniform procurement and management involves not only selecting more sustainable materials but also considering what happens to garments after their initial use. This focus on both life cycle extension and responsible end-of-life planning forms a crucial part of sustainability strategies across the industry (Streamline, 2023).

Benefits of sustainable uniforms

- **Environmental impact:** Conventional uniforms are commonly made from synthetic fibres like polyester, which are derived from petroleum and associated with high carbon emissions. In contrast, sustainable alternatives – such as organic cotton, bamboo or recycled polyester – require less energy and water during production and often break down more easily in landfill or can be recycled into new materials (Streamline, 2023).
- **Employee health and safety:** Sustainable uniforms are often manufactured without the use of toxic chemicals, pesticides or dyes. This reduces the potential for skin irritation, allergic reactions and other health concerns for employees who wear the garments for extended periods (Streamline, 2023).
- **Cost and durability:** Although the upfront cost of sustainable uniforms can be higher, they are generally more durable and require less frequent replacement. Some suppliers also offer garment return or recycling programmes, which can reduce disposal costs and keep textiles out of landfill (Streamline, 2023).
- **Corporate social responsibility:** Sourcing uniforms from ethical and environmentally responsible suppliers supports transparency in the supply chain. It also reflects positively on an organization's values, particularly in areas such as fair labour practices and environmental stewardship, which are increasingly important to consumers and stakeholders.

Strategies for extending uniform life cycles

Adopting circular design and procurement practices can help reduce waste and extend the useful life of hospitality uniforms. Key strategies include:

- **Material selection:** Uniforms made from organic or recycled fibres tend to have lower environmental footprints and can often be recycled or biodegraded more easily.
- **Ethical sourcing:** Working with certified suppliers that follow fair labour practices ensures ethical standards throughout the supply chain. Recognized certifications include Fairtrade, WRAP (Worldwide Responsible Accredited Production) and Sedex.
- **Design for longevity and adaptability:** Uniforms can be designed to serve multiple roles or be easily adjusted, allowing them to remain in use longer and be repurposed as staffing needs change.
- **Recycling and reuse initiatives:** Businesses can implement systems for collecting and reusing old uniforms, or for recycling fabric into new garments or industrial materials.
- **Maintenance education:** Providing employees with guidance on caring for uniforms – such as washing in cold water or air drying – can help extend garment lifespan and reduce environmental impact.
- **Local sourcing:** Procuring uniforms from local suppliers can reduce transportation emissions and support domestic manufacturing industries.

Incorporating sustainable uniform practices into hospitality operations aligns with circular economy principles by reducing waste, improving resource efficiency and extending the life cycle of products. While such changes may begin with modest adjustments, they contribute to broader environmental goals and long-term operational resilience.

End-of-life management

When products reach the end of their life cycle, proper management ensures that waste is minimized, and materials are recovered. Regulations such as the UK's Extended Producer Responsibility mandate businesses to design products that can be reused or recycled (DEFRA, 2022). The aim of this type of regulation is to ensure that manufacturers are held accountable for the recovery and recycling targets associated with the product. These measures aim to restrict the environmental impacts of a throwaway culture, where items are discarded and replaced rather than repaired or repurposed.

For the THE industry, the responsibility lies in raising awareness among staff and customers about the importance of extending product life cycles and adopting proper disposal practices. By realigning consumer and operational behaviours, such efforts promote responsible consumption and reduce the frequency of discarding goods.

Extending product life and managing end-of-life responsibly are interconnected strategies. While the former maximizes a product's usable life, the latter ensures responsible recovery or disposal. When combined, these practices reduce the volume of waste generated, minimize resource extraction and support a more efficient, closed-loop system.

By integrating circular economy principles of sustainable production, responsible consumption, resource optimization and effective waste management – including extending product life cycles and end-of-life practices – THE can significantly reduce its environmental footprint and carbon emissions. However, the transition to a circular economy is not the responsibility of individual businesses alone. It requires collective effort and collaboration between industries, policy makers and communities to create shared solutions. Cross-industry collaboration and circularity describes how THE industries can work together to drive the transition towards a more interconnected and circular economic system highlighting the importance of partnerships, transparency, and cooperative policy and practice.

Cross-industry collaboration and circularity

A fundamental aspect of the circular economy is interconnectedness – no single industry within the THE sector can achieve sustainability in isolation. By implementing unique sustainability initiatives and sharing best practice, industry leaders can share their knowledge and resources to create scalable solutions (World Sustainable Hospitality Alliance, 2025a). Collaboration allows for the development of sustainability practices that ensure that positive change is not just limited to a few organizations, but can be implemented across the entire industry (World Sustainable Hospitality Alliance, 2025b). Collaboration and transparency are essential for enabling circularity within the THE supply chain. By developing partnerships among diverse stakeholders including venues, suppliers, policymakers, and local communities, the industry can transition from a linear resource model to a more sustainable and circular economy.

Collective industry action and collaboration

Collaboration among industry leaders is central for creating scalable solutions that benefit the environment and the THE sector. Non-competitive, collective efforts enable the sharing of knowledge and resources, making sustainability practices more accessible and affordable across the industry. A critical element of collaboration is leveraging group purchasing power, which allows businesses to lower the costs of sustainable products and practices. This approach particularly benefits small and medium-sized enterprises (SMEs), enabling them to compete with larger chains in offering sustainable solutions while accelerating the adoption of circular economy principles (World Sustainable Hospitality Alliance, 2025b).

By pooling resources, THE operations streamline the sourcing process, reducing reliance on multiple suppliers and enhancing efficiency. Group action also encourages innovation, allowing businesses to co-create solutions such as recyclable products, eco-friendly amenities and waste reduction strategies that align with sustainability goals.

Building a sustainable and transparent system

Transparency plays a vital role in building trust and accountability within the supply chain. By openly sharing information about product origins, carbon footprints and ethical sourcing, businesses ensure that sustainability claims are credible and traceable. For instance, providing detailed information about sustainably sourced linens or biodegradable toiletries enables guests and stakeholders to make informed decisions (World Sustainable Hospitality Alliance, 2025b).

The development of feedback loops is another essential aspect of transparency (World Sustainable Hospitality Alliance, 2025b). Continuous improvement is achieved as stakeholders exchange insights on product performance, recyclability and waste management. Businesses can report challenges to suppliers, who can refine their offerings to align with circular economy principles.

By incorporating continuous improvement and feedback loops, any collaborative initiative will ensure that it remains responsive to real-world needs and is constantly adapting to new sustainability challenges.

Supporting small suppliers and local communities

Supporting small suppliers is a critical component of collaboration across THE. By connecting small, local businesses with larger brands, the industry creates a diverse and inclusive supply chain that empowers local communities and engenders innovation in sustainable products. Small suppliers, who often face challenges in scaling operations, gain access to wider markets and larger contracts, enabling them to grow while promoting ethical sourcing (World Sustainable Hospitality Alliance, 2025a).

The development of local supply chains contributes to regional economic growth and strengthens the environmental sustainability of the industry. Sourcing organic produce from local farmers, for example, minimizes transportation emissions and creates a closed-loop system that benefits both the community and the environment.

Figure 6.6 Sustainability initiatives undertaken by the organizers of Glastonbury Festival

Glastonbury Festival Sustainability Metrics

Key Festival Facts
- Annual attendance: Over 210,000 people
- Founded in 1970
- Break every 5 years to allow ground recovery
- Located at Worthy Farm, a working farm

Energy & Renewable Initiatives
- Powered entirely by renewable energy as of 2023
- 250kWp solar PV array
- 125kVa biogas plant
- 28-metre wind turbine (300kW/day)
- HVO fuel from waste cooking oil for all generators
- Solar panels save 100.9 tonnes of CO_2 annually

Waste Management
- 2,000 tonnes of waste generated yearly

Recycled since 2019:
- 68 tonnes of paper and card
- 38 tonnes of glass
- 57 tonnes of cans
- 17 tonnes of plastic bottles
- 14,000 litres of cooking oil turned into biofuels
- 98% of tents taken home in 2023

Plastic Reduction
- 2019: Single-use plastic bottles banned
- 2022: Non-compostable crisp packets banned
- 2023: Disposable vapes banned
- Only compostable or reusable serveware allowed

Water Conservation
- Bristol Water supplies over 800 water points
- Auto shut-off fixtures on taps
- 700m of urinals, 2,000 long-drop toilets, 1,200 compost toilets

Tree Planting & Environmental Impact
- Over 10,000 native trees and hedges planted since 2000
- 175,700kWh of renewable energy generated annually

Data from Birch, 2024

Enhancing the guest experience

Sustainable practices not only benefit the environment but also enhance the guest experience across tourism, hospitality and events. By offering eco-friendly amenities and visibly demonstrating their commitment to sustainability, THE operators and organizers align with the growing demand for environmentally responsible travel and experiences. Guests are empowered to participate in sustainable practices, such as recycling programmes, water conservation initiatives and waste reduction strategies at hotels, tourist attractions and event venues. This active engagement encourages a shared sense of responsibility between businesses and customers (World Sustainable Hospitality Alliance, 2025a).

Visible metrics can play a significant role in this process. For example, a festival such as Glastonbury in the UK attracts more than 210,000 attendees per year. While it produces more than 2,000 tonnes of waste every year, it also saves nearly 6,000 tonnes of greenhouse emissions (Birch, 2024). Figure 4.5 presents data illustrating the level of sustainability of Glastonbury Festival (Birch, 2024). It shows how complex metrics can be presented in a consumer-friendly way to not only educate but also engage consumers and encourage considerations of sustainable behaviour.

The aim of collaboration for a circular economy is to achieve 'net positive' THE, where the industry not only reduces its environmental footprint but also creates positive environmental and social outcomes (World Sustainable Hospitality Alliance, 2025a). A good example of the circular economy in this sense is the Community Homestay Network.

REAL-WORLD EXAMPLE The Community Homestay Network (CHN): An example of collaboration and circularity

Achieving net zero in a circular economy context involves collaboration across industries, sectors and communities. Stakeholders must work together to develop solutions that reduce emissions and promote resource efficiency. For example, a five-star hotel that engages in partnerships with companies to promote bike tours is encouraging eco-friendly travel options for tourists. These initiatives help reduce the overall carbon footprint of the destination while supporting the local economy.

The Community Homestay Network (CHN) serves as a good illustration of the principles of a circular economy that not only addresses the environmental but extends to incorporate social dimensions. The CHN integrates collaboration and circularity into a broader framework of responsible tourism, using homestays as a foundation for empowering local communities, preserving cultural heritage and promoting environmental sustainability. For example, the CHN operating in Nepal involves the Nepal Tourism Board partnering with local businesses to promote sustainable tourism (Community Homestay Network, .n.d. a). Their circular economy focuses on the regeneration of resources, optimization of systems and minimization of waste.

The following principles driving CHN demonstrate how THE can take a circular approach to its operations:

- **Retaining value locally**: Revenue generated from homestays and community experiences remains within the local economy, supporting families, artisans, and businesses.
- **Regenerating community assets**: Reinvestment in communities by providing training, enhancing infrastructure and promoting cultural preservation creates a self-sustaining system of improvement.
- **Minimizing external costs**: Eco-tourism practices and cultural conservation ensures that tourism strengthens local environments and traditions rather than depleting them (Community Homestay Network, 2025).

Their circular model is operationalized through emphasis on community ownership, collaboration and sustainable practices, creating an ecosystem where tourism generates economic, social and environmental benefits simultaneously. By collaborating with local communities to design and manage tourism offerings local women, youth, and marginalized groups are also being empowered (Community Homestay Network, .n.d. b). As CHN say, 'By collaborating with the communities and various impact partners, CHN helps women break biases and engage in economic activities, leading to greater social empowerment and cultural awareness ... This comprehensive training helps women effectively interact with guests, exchange ideas and culture, and manage tourism activities, ensuring communities are well-equipped to offer authentic and responsible tourism experiences.' Community Homestay Network, .n.d. b)

Through collaboration, transparency and a commitment to circularity, THE not only minimizes waste and enhances resource efficiency, but also builds trust and accountability across the supply chain. As the industry continues to evolve, these principles will remain fundamental to achieving a sustainable and circular future for THE (World Sustainable Hospitality Alliance 2025a; 2025b).

ACTIVITY 1: CIRCULAR ECONOMY ON THE RIVER: ENHANCING THE GUEST EXPERIENCE THROUGH SUSTAINABILITY

Serenity Sailing is a river cruise company committed to minimizing its environmental footprint and maximizing guest satisfaction. They are currently implementing circular economy principles into their operations and want to share these initiatives with their guests.

Serenity Sailing believe that their river cruises offer a unique setting to connect travellers with sustainability in a tangible and engaging way.

Purpose

The objectives are to:

- understand the principles of a circular economy and its relevance to THE
- analyze how circular economy principles can be applied to enhance the guest experience on a river cruise
- develop innovative strategies to educate and engage guests in sustainable practices
- design a circular economy model for a river cruise that prioritizes both environmental responsibility and guest satisfaction
- evaluate the potential impact of these strategies on the river cruise company, guests and the environment

Instructions

Serenity Sailing have asked you to develop a proposal on how they can enhance the guest experience by providing insightful and entertaining activities for all ages on their circular activities. The proposal should outline how they can introduce their guests to circular economy principles with a strong emphasis on an engaging and insightful guest engagement. They also want you to include ideas on how to measure the success of the engagement activities.

Your proposal should design a guest experience programme that considers the following areas:

1 **Enhancing the onboard experience through circularity**
   o Identify opportunities to integrate circular economy principles into various aspects of onboard guest experience, such as:
     – **Dining**: Design menus that feature locally sourced, seasonal ingredients and minimize food waste. Explore closed-loop systems in operation on the ship (e.g. composting food waste for onboard gardening).

- **Amenities**: Propose strategies to reduce single-use plastics and packaging in cabins and public areas.
- **Activities**: Develop onboard activities and shore excursions that highlight circular economy principles in action
  - Explain how these changes can enhance guest satisfaction and create a unique selling proposition for Serenity Sailing.

2 **Communication and storytelling**
  - Develop a compelling narrative that communicates Serenity Sailing's commitment to the circular economy to its guests.
  - Identify the most effective channels to communicates these initiatives (e.g. website, social media, onboard announcement, printed materials).
  - Create engaging content that showcases the positive impact of the circular economy initiatives on the environment and local communities.
  - Train staff to effectively communicate the company's circular economy vision and answer guest questions.

3 **Partnerships and collaborations**
  - Identify potential partners that could collaborate with Serenity Sailing to enhance guest engagement for the benefit of the local community.
  - Propose ways to showcase these partnerships to guests and highlight the positive impact of collaboration.

Your proposal to Serenity Sailing should include ideas for onboard presentations, interactive workshops, digital content, guided tours and activities for children to cater for a diversity of guest ages and learning styles. It should evidence how to develop engaging content that connects circular economy concepts to the river cruise experience, highlighting local ecosystems, cultural heritage and sustainable practices in the destinations visited.

4 **Measuring the impact on the guest experience**
  - Develop a framework for measuring how the proposed initiatives affect guest satisfaction.
  - Identify key performance indicators (KPIs) related to the guest experience, such as:
    - guest feedback on onboard activities and educational opportunities
    - satisfaction with dining and accommodation
    - perceptions of Serenity Sailing's sustainability efforts
  - Propose methods for collecting and analyzing data on guest satisfaction (e.g. surveys, feedback forms, online reviews).

Conclusion

In this chapter, we have explored how the circular economy offers a practical and innovative approach to sustainability within THE industries. As THE faces growing pressure to respond to climate change, cut greenhouse gas emissions and support global sustainability goals, the circular economy emerges as a compelling alternative to the outdated 'take, make, dispose' model.

Through an examination of circular principles and real-world applications – Hotel Casa Palmela in Portugal and the Community Homestay Network (CHN) in Nepal – we have seen how THE businesses can implement closed-loop systems, reduce environmental impacts and enhance guest experiences through innovative, circular practices.

Key to this transition are the UNESCO sustainability competencies: systems thinking, critical thinking, strategic action and collaboration. These skills are essential for analysing complex sustainability challenges, designing creative solutions and working with diverse stakeholders to co-create resilient and regenerative experiences. By embedding circular economy thinking into both strategy and everyday operations, THE professionals can contribute to a more sustainable, inclusive and climate-positive future.

> **KEY TAKEAWAYS**
>
> - The circular economy:
>   - The circular economy provides a viable and future-focused approach to reducing waste, conserving resources, and building more sustainable THE operations.
>   - THE businesses can apply circular practices to enhance efficiency and improve the environmental and economic value of their services.
>   - Real-world examples like Hotel Casa Palmela and the Community Homestay Network illustrate how circularity can be embedded into luxury and community-based tourism models, offering best practice for innovation across the sector.
>   - Thinking in systems and challenging conventional practices enables professionals to identify opportunities for circular transformation in design, operations and guest experiences.
>   - Effective circular strategies require collaboration, creativity and practical problem-solving, making teamwork and interdisciplinary thinking essential for success in THE industries.

> **REFLECTIVE QUESTIONS**
>
> 1 Reflecting upon the proposal you developed for the river cruise company, what aspects of the circular economy do you think would most effectively engage guests, and why? How could these initiatives be scaled or adapted across different types of hospitality settings?
>
> 2 Based upon the hotel real-world example you examined, what do you think were the key enablers and barriers to implementing circular economy principles? How might similar strategies be applied – or improved – within other tourism or events-based environments?
>
> 3 Looking ahead, how could you apply what you have learned about circular economy practices in your future professional role within THE? What practical actions would you prioritize?

References

Amicarelli, V, Aluculesei, A-C, Lagioia, G, Pamfilie, R. and Bux, C (2022) How to manage and minimize food waste in the hotel industry: an exploratory research, International *Journal of Culture, Tourism and Hospitality Research*, 16 (1), pp 152–67, https://doi.org/10.1108/IJCTHR-01-2021-0019 (archived at https://perma.cc/G2T8-GCMP)

Association of Plastic Recyclers (2018) 8 lifecycle impacts for postconsumer recycled resins: PET, HDPE, and PP, https://plasticsrecycling.org/images/library/2018-APR-LCI-report.pdf (archived at https://perma.cc/MU3X-F6YW)

Baloch, Q B, Shah, S N, Iqbal, N, Sheeraz, M, Asadullah, M, Maha, S and Khan, A U (2022) Impact of tourism development upon environmental sustainability: a suggested framework for sustainable ecotourism, *Environmental Science and Pollution Research*, 30, pp 5917–30, https://doi.org/10.1007/s11356-022-22496-w (archived at https://perma.cc/V6NC-UADU)

Birch, T (2025) How sustainable is Glastonbury Festival? The Eco Experts, 17 June, www.theecoexperts.co.uk/news/glastonbury-festival-carbon-footprint (archived at https://perma.cc/3UL9-XYHR)

Broom, D (2023) What are the world's biggest natural carbon sinks? World Economic Forum, 26 July, www.weforum.org/stories/2023/07/carbon-sinks-fight-climate-crisis (archived at https://perma.cc/23U6-73T4)

Casa Palmela Hotel, Arrábida (.n.d.) https://hotelcasapalmela.pt (archived at https://perma.cc/2TQJ-7NJJ)

Community Homestay Network (2025) Community Connect 2025: Unveiling Nepal's hidden gems through responsible tourism, 7 April, https://communityhomestay.com/blog/community-connect-2025 (archived at https://perma.cc/PR8Y-QNEN)

Community Homestay Network (.n.d. a) About Us, https://communityhomestay.com/about (archived at https://perma.cc/97PV-V7PJ)

Community Homestay Network (.n.d. b) What does women empowerment mean for the communities? https://communityhomestay.com/impact (archived at https://perma.cc/B4LF-ECP4)

DEFRA (2022) Extended producer responsibility for packaging: who is affected and what to do, GOV.UK, 7 June, www.gov.uk/guidance/extended-producer-responsibility-for-packaging-who-is-affected-and-what-to-do (archived at https://perma.cc/TC74-SKG3)

Ellen MacArthur Foundation (2021) Completing the picture: How the circular economy tackles climate change, 26 May, https://ellenmacarthurfoundation.org/completing-the-picture (archived at https://perma.cc/WE5W-9EX3)

Environmental Literacy Council (2024) How do landfills produce methane? 4 October, https://enviroliteracy.org/how-do-landfills-produce-methane (archived at https://perma.cc/QA73-ELCP)

Fairtrade (n.d.) www.fairtrade.org.uk/ (archived at https://perma.cc/43CT-V8PL)

International Air Transport Association (.n.d.) Non-CO2 Emissions – Contrails, www.iata.org/en/programmes/sustainability/non-co2-emissions-contrails (archived at https://perma.cc/7FBU-EAS5)

IPCC (2023) Climate Change 2023 : AR6 Synthesis Report: longer report, UN Digital Library, https://digitallibrary.un.org/record/4008074?v=pdf (archived at https://perma.cc/KT3M-83DG)

Knight, C (2023) What is the linear economy? European Investment Bank, 2 August, www.eib.org/en/stories/linear-economy-recycling (archived at https://perma.cc/XV2U-NUJH)

legislation.gov.uk (2011) The Waste (England and Wales) Regulations 2011, www.legislation.gov.uk/uksi/2011/988/contents (archived at https://perma.cc/6BAP-BJ7V)

legislation.gov.uk (2024) Circular Economy (Scotland) Act 2024 (asp 13), www.legislation.gov.uk/asp/2024/13/pdfs/asp_20240013_en.pdf (archived at https://perma.cc/9EBW-C88V)

Lenzen, M, Sun, Y-Y, Faturay, F, Ting, Y-P, Geschke, A and Malik, A (2018) The carbon footprint of global tourism, *Nature Climate Change*, 8 (6), pp 522–28, www.nature.com/articles/s41558-018-0141-x (archived at https://perma.cc/QV2W-G2VG)

Manniche, J, Topsø Larsen, K, Brandt Broegaard, R and Holland, E (2018) Destination: A circular tourism economy. A handbook for transitioning toward a circular economy within the tourism and hospitality sectors in the South Baltic region, Researchgate, November, www.researchgate.net/publication/337085400_Destination_A_circular_tourism_economy_A_handbook_for_transitioning_toward_a_circular_economy_within_the_tourism_and_hospitality_sectors_in_the_South_Baltic_Region_Final_edition (archived at https://perma.cc/42DW-UKKP)

NASA (.n.d.) What is the greenhouse effect? https://science.nasa.gov/climate-change/faq/what-is-the-greenhouse-effect (archived at https://perma.cc/Z6HG-D7G3)

Pereira, V, Silva, G and Dias, Á (2021) Sustainability practices in hospitality: Case study of a luxury hotel in Arrabida Natural Park, *Sustainability*, 13 (6), p 3164, https://doi.org/10.3390/su13063164 (archived at https://perma.cc/YMF4-Z7FN)

Streamline (2023) Sustainable work uniforms: How small changes can create a big impact for your team, 24 October, www.brandedbystreamline.com/sustainable-work-uniforms-how-small-changes-can-create-a-big-impact-for-your-team (archived at https://perma.cc/QV6N-PKMN)

Too Good To Go ApS (2024) Impact Report 2024, www.toogoodtogo.com/impact-report (archived at https://perma.cc/KQ9N-9U25)

Unegg, M C, Steininger, K.W, Ramsauer, C and Rivera-Aguilar, M (2023) Assessing the environmental impact of waste management: A comparative study of CO_2 emissions with a focus on recycling and incineration, *Journal of Cleaner Production*, **415**, pp 1–10, https://doi.org/10.1016/j.jclepro.2023.137745 (archived at https://perma.cc/AEZ4-HTZT)

UNEP (2024) Emissions Gap Report 2024: No more hot air... please! www.unep.org/resources/emissions-gap-report-2024 (archived at https://perma.cc/E26R-N7EN)

UNESCO (2017) Education for Sustainable Development Goals Learning Objectives, https://unesdoc.unesco.org/ark:/48223/pf0000247444 (archived at https://perma.cc/VVZ2-8KW8)

UNWTO (2021) Glasgow Declaration: Climate Action in Tourism, www.unwto.org/glasgow-declaration-climate-action-in-tourism (archived at https://perma.cc/CV8J-ZVWC)

World Sustainable Hospitality Alliance (2025a) Net positive goals for 2025: nature fluency & decarbonising food systems, 10 February, https://sustainablehospitalityalliance.org/net-positive-hospitality-podcast (archived at https://perma.cc/NVE8-UB5Z)

World Sustainable Hospitality Alliance (2025b) The planet conscious amenities pack: Revolutionising hotel sustainability, 27 February, https://sustainablehospitalityalliance.org/net-positive-hospitality-podcast (archived at https://perma.cc/2XBX-7HYH)

7 | Food waste

| Systems thinking competency | Critical thinking competency | Strategic competency | Integrated problem-solving competency |

CHAPTER AIM

This chapter aims to provide a comprehensive understanding of food waste within the tourism, hospitality and events (THE) sector. It explores key definitions, the global scale of the issue and its significant environmental, social and economic impacts. Emphasis is placed on aligning food waste reduction efforts with key global frameworks, including Sustainable Development Goals (SDGs) like SDG 12, SDG 2, SDG 13, and SDG 15.

The chapter focuses on developing practical solutions for prevention and management, supporting key sustainability competencies such as systems thinking, strategic competency, critical thinking and integrated problem-solving. Through examining the food supply chain, specific interventions and best practices, learners are equipped to address food waste effectively in operational contexts.

LEARNING OUTCOMES

Upon completion of this chapter, you will be able to:

- define food waste and food loss and explain the global scale and multi-faceted impacts of food waste
- analyze the food supply chain to identify key stages and causes of food loss and waste
- evaluate various practical interventions and strategies for preventing food waste in THE operations
- describe methods for managing unavoidable food waste sustainably (e.g. composting, anaerobic digestion)
- recognize the role of human factors (consumer and staff behaviour) in food waste generation and reduction
- identify relevant initiatives and best practices for food waste management in the THE sector

KEY WORDS

Food waste, food loss, supply chain, carbon footprint, greenhouse gas emissions (GHG), waste prevention, waste management, circular economy.

Introduction

In a world grappling with climate change and resource scarcity, food waste presents a critical challenge with profound environmental, economic and social consequences, particularly within the global tourism, hospitality and events (THE) sector. While millions face hunger, staggering amounts of edible food are discarded annually.

This chapter explores the multifaceted issue of food waste. It begins by establishing key definitions of food waste and food loss and examining the immense scale of the problem globally and within the UK, including its significant carbon, water and land footprints. The chapter then analyzes food waste in the supply chain before exploring practical solutions, focusing on prevention strategies for operations and sustainable management techniques such as composting and anaerobic digestion. Next, the chapter considers the crucial role of human behaviour and highlights best practices and initiatives designed to combat food waste effectively (FAO, 2022).

Finally, the chapter presents a real-world example illustrating best practices in food waste management within a particular operational setting, followed by a student activity designed to help learners synthesize the key takeaways of Chapter 7.

Definition of food waste

This section defines food waste, considering both geographical location and its place within the food system. According to the Food and Agriculture Organization (FAO), 2022, 'food' encompasses everything intended for human consumption, including inedible parts like pineapple rinds. 'Food loss' refers to unintentional quality or quantity deterioration during production, storage, processing and distribution. Spoilage, spills and infrastructure limitations can all contribute to food loss. In contrast, 'food waste' according to the World Trade Organization (UNWTO, 2024), indicates any food or inedible part diverted from the food chain for recovery or disposal. This encompasses composting, incineration, landfills, and even dumping at sea. It also includes anaerobic digestion, a process where microorganisms break down organic material like food scraps in an oxygen-free environment to produce biogas (for energy) and digestate (a soil conditioner). The rationale behind this broad definition is that food waste, even inedible parts, holds valuable resources like carbon, water and nutrients. Recovering and recycling this waste offers environmental benefits.

In certain international contexts (such as by the FAO) the term 'food wastage' is also used. This refers to any food lost by deterioration or waste. Thus, the term 'wastage' according to the FAO (2024) encompasses both food loss and food waste. However, for clarity's sake and in line with mainstream terminology the term food waste used in this book also encompasses 'food wastage'.

A further differentiation can be made between 'avoidable food waste,' which was once edible (bread, apples, meat), and 'unavoidable food waste' generated during food preparation, such as bones and eggshells. Within cities, food waste primarily originates from households, restaurants, processing plants, shops and storage facilities, not farms or fisheries. The consequences of food waste extend far beyond the depletion of resources. Poorly managed food waste contributes to climate change through greenhouse gas emissions during decomposition, pollutes water sources through nutrient runoff and can even become a breeding ground for diseases (WRAP, 2024).

Food waste: The scale of a global issue

The FAO has measured the impact of food waste on natural resources, particularly its carbon footprint – a measure which dates back to 2011 (see Figure 7.1 above).

Figure 7.1 Total carbon footprint

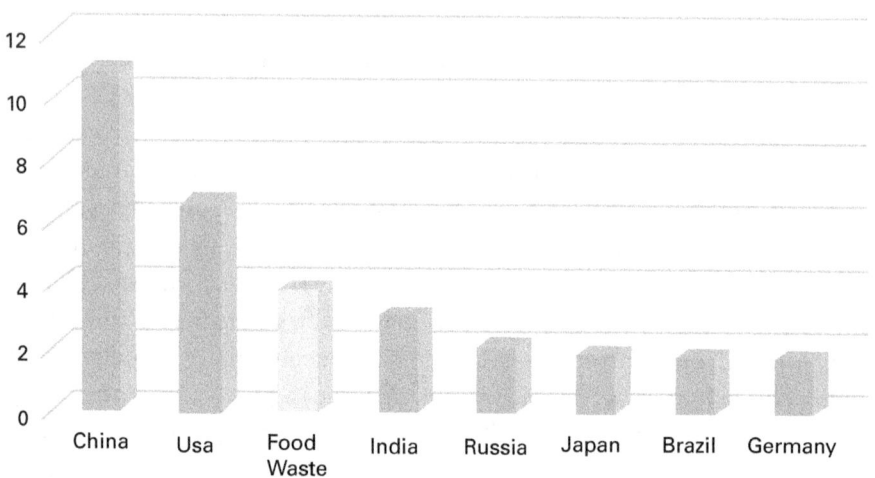

Adapted from Guardians of Grub, 2023

The carbon footprint of any food item represents the total greenhouse gas (GHG) emissions produced throughout its life cycle. To compare the warming impact of different GHGs (like methane or nitrous oxide) against that of carbon dioxide (CO_2), their emissions are commonly quantified in terms of CO_2-equivalence (CO_2 eq). This standard unit expresses the global warming potential of each gas relative to one unit of CO_2, typically over a 100-year period. For instance, over this timeframe, one tonne of methane has a global warming potential equivalent to many tonnes of CO_2.

When discussing very large volumes of emissions, such as those at a global scale, the unit 'gigatonnes of CO_2 equivalent' ($GtCO_2$ eq) is often used. One gigatonne (Gt) is equal to one billion metric tonnes. Thus, $GtCO_2$ eq provides a way to express massive quantities of mixed greenhouse gas emissions in a single, comparable figure. This comprehensive calculation typically includes emissions throughout the entire supply chain, from agricultural inputs and land use change through to processing, transportation, consumption and waste disposal.

Food waste volumes and emission factors derived from life cycle assessment studies are estimated to be 3.6 $GtCO_2$ eq, which does not include the 0.8 $GtCO_2$ eq from deforestation and managed organic soils linked to food wastage. Therefore, the total carbon footprint of food waste, including changes in land use, is approximately 4.4

GtCO$_2$ eq per year. The annual global food loss and waste generate 4.4 GtCO$_2$ or about 85 per cent of anthropogenic GHG emissions. This implies that the contribution of food waste emissions to global warming is nearly equivalent to international road transport emissions (IPCC, 2022).

Food waste represents a significant inefficiency in the market, leading to the disposal of food valued at over US$1 trillion annually (UNEP, 2024). This reflects an economic and environmental shortfall: the production and disposal of wasted food contribute to 8–10 per cent of worldwide greenhouse gas emissions. Moreover, it occupies nearly a third of the planet's agricultural land. The transformation of natural habitats for farming purposes is the primary driver of habitat destruction. The human impact of food waste is equally pressing: while substantial quantities of food are discarded, hunger affects up to 783 million individuals annually. Additionally, due to insufficient intake of vital nutrients, stunted growth and development afflict 150 million children under five years old.

This situation underscores the urgent need to address food waste from a sustainability perspective in and in accordance to our field within the THE sectors. For instance, a kilogram of wheat and a kilogram of beef have distinct carbon footprints due to their differing life cycles, which emit varying types and amounts of greenhouse gases. The carbon intensity of products also varies. Vegetable production in Europe, for example, is more carbon-intensive than in industrialized Southeast Asia due to the use of more carbon-intensive production methods, such as artificially heated greenhouses. Conversely, cereal production in Asia is more carbon-intensive than in Europe due to the type of cereal grown: rice, on average, has a higher impact factor than wheat (Yang et al, 2024). This is because rice cultivation emits methane due to the decomposition of organic matter in paddy fields (1kg of methane is equivalent to 85 kg of CO$_2$).

Despite meat contributing less than 5 per cent to global food wastage volumes, it significantly impacts climate change, accounting for over 20 per cent of the total food waste carbon footprint. This is because the carbon footprint of meat includes emissions from producing a kilogram of meat (e.g. methane emitted by ruminants during digestion), emissions related to feed provision (e.g. the cultivation of feed crops often requires fertilizers, which themselves have a carbon footprint, and can involve land-use change), and emissions from manure management (Symeon et al, 2025). Therefore, given their disproportionately high environmental impact relative to their production volume or waste quantity, efforts to reduce GHG emissions related to food wastage should strategically prioritize these major 'climate hotspot' commodities, such as meat and cereals. Meat, as detailed, has a large carbon footprint per kilogram due to resource-intensive production processes. Similarly, certain cereals, particularly rice grown in paddy systems, can be significant GHG contributors due to methane emissions from anaerobic decomposition during cultivation. Focusing efforts on such commodities can thus yield more substantial reductions in the overall GHG emissions stemming from the food system.

Understanding the dynamics of food waste sources is crucial. Food loss and waste occur at various stages of the supply chain journey: production, processing, distribution, retail, and consumption. In developing countries, the majority of food loss happens during the production, processing and distribution stages. This is largely due to infrastructural challenges, such as inadequate storage facilities (leading to pest infestation or spoilage), limited access to refrigeration, and inefficient transport and supply chains that can result in damage or deterioration before food reaches the market.

On the other hand, in developed countries, food waste is more prevalent at the retail and consumption stages. This is often driven by consumer behaviour and societal expectations around food aesthetics and freshness. For instance, perfectly edible food items might be discarded simply because they do not meet stringent cosmetic standards for shape, size or colour. According to UNEP (2024), on a per capita basis, global consumers on average produce 74 kg of food waste annually, with more recent statistics pointing to 77 kg (UNEP, 2024). The picture in Europe, according to Eurostat (2023), shows an estimated 131 kilograms of food waste generated per person in the EU. Of this, households were responsible for 54 per cent, equating to 70 kg per person. The remaining 46 per cent was waste produced higher up in the food supply chain.

The amount of food waste from households is nearly double that from the sectors of primary production (e.g. farming and fishing) and the manufacturing of food products and beverages (11 kg and 28 kg per person, or 9 per cent and 21 per cent respectively). This disparity can be partly attributed to the fact that these industrial sectors often have established strategies to minimize what is formally classified as 'waste' by finding alternative uses for materials that are not part of the primary product, thereby enhancing resource efficiency. For example, vegetable trimmings might be processed into soup bases, animal feed or biofuels; fruit peels could be used for pectin extraction, essential oils or composting; and byproducts from grain milling (like bran or germ) can be incorporated into other food products or animal nutrition. This approach, often aligned with circular economy principles, aims to extract maximum value from agricultural outputs and industrial processes, thereby reducing the final volume of materials that are simply discarded. Restaurants and food services contributed 12 kg of food waste per person (9 per cent), while the retail and other food distribution sectors generated the least amount of food waste among these categories (9 kg or 7 per cent) (WRAP, 2024).

In developing nations, the ratio of 'food waste' (typically occurring at consumer or retail level) to 'food loss' (occurring earlier in the supply chain) is significantly smaller. The primary cause of food loss in these regions, as mentioned, is inefficiencies during the stages of agricultural production, post-harvest handling, storage and processing. These inefficiencies can be attributed to factors such as premature harvesting, inadequate drying or curing techniques, lack of access to proper storage facilities (leading to spoilage from heat, moisture or pests), limited processing capabilities, and insufficient infrastructure for transportation and cold chain management, In developing nations, there tends to be also underdeveloped market systems that may not be able to absorb all produce. Food waste that does occur in developing countries at the consumption end is often primarily made up of inedible parts of food, like peels, shells and pulp. These remnants might be what is left after human consumption or could be a byproduct or waste from the food and beverage industry's processing activities.

Contrastingly, in industrialized nations, there is a notable increase in waste and losses during the distribution and consumption stages. Food that could still be consumed is often discarded for a variety of reasons. These multifaceted factors, which contribute significantly to the food waste challenge in wealthier nations – including stringent retail cosmetic standards, consumer preferences for aesthetically perfect items, confusion over date labelling (such as 'best before' versus 'use by' dates), over-purchasing by households, improper food storage at home and inadequate meal planning – will be explored in more detail later in this chapter when discussing specific causes and intervention points.

Global impact of food waste on natural resources

GHG emissions resulting from changes in land use linked to food production, such as the deforestation of the Amazon rainforest for additional farmland, significantly increase the global carbon footprint of food waste. However, this category of emissions is challenging to quantify. Excluding land use change, it is noteworthy to remember that if the equivalent of the carbon footprint of food waste were a country, it would be the third-largest emitter of GHGs in the world, trailing only the USA and China (see Figure 7.1 above).

Another footprint created by food waste is the 'blue water footprint'. The blue water footprint refers to the consumption of surface water and groundwater used to produce food that is ultimately uneaten. It represents wasted irrigation water from rivers, lakes and natural underground reservoirs (aquifers) (Amicarelli, Lagioia and Bux, 2021). The global blue water footprint of food waste, is approximately 250 km^3 (Amicarelli, Lagioia and Bux, 2021). The blue water footprint of food wastage surpasses any country's blue water footprint for global crop production wasting 174km^3 of blue water (Amicarelli, Lagioia and Bux, 2021).

The global footprint of food waste, which refers to the total land area used to cultivate food that ultimately gets wasted, represents an area larger than either Canada or China and is only surpassed by the size of the Russian Federation. It is worth mentioning that a significant portion of food waste at the agricultural production stage tends to occur in regions where the soil is undergoing moderate to severe degradation. These regions are often the poorest, where a cycle of land degradation threatens the food security of the most vulnerable populations.

The biodiversity footprint of food waste is also substantial. Agriculture, including land conversion and intensification, poses a significant threat to global biodiversity. The threats are primarily due to crop production rather than livestock production. In both cases, biodiversity loss is considerably greater in Latin America, Asia (excluding Japan) and Africa than in Europe, Oceania, Canada and the USA. This could be partially explained by the fact that tropical countries have environments rich in biodiversity, regardless of management intensity. In addition to its footprints, food waste carries both a financial and a social cost, not to mention its contribution to global hunger. Besides the monetary value of the food itself (i.e. the value of the product at the stage during which it was wasted), the natural resources embedded in the wasted food also hold value. Given the increasing scarcity of global resources, such as land and water, the price of natural resources is expected to rise in the future. In many countries, water and land already carry high costs, and GHG emissions lead to climate changes which can have significant economic implications (UNEP, 2016).

The high social cost is due to food waste depleting resources on which the poorest are most dependent. If wealthier countries wasted less water and other limited resources as part of food waste, it would free up agricultural land and other resources to grow food crops such as cereals, that could contribute to much needed global supplies. This sequence is most evident for internationally traded commodities such as wheat, and less obvious, but still applicable, for fresh produce grown and purchased within individual nations. Moreover, wasting food in wealthier countries directly contributes to global hunger. Whether wealthy or poor, all countries purchase food from the same global market of internationally traded commodities. If wealthier countries buy hundreds of millions of tonnes of food they end up wasting, they are removing food from the market which could have remained there for other countries to buy. By increasing demand for these commodities, wealthier countries also contribute to price increases, which makes them less affordable for poorer nations.

When food waste takes place at a specific stage of the food supply chain (see Figure 7.2 below), three types of impacts need to be taken into account: the impact on the production phase itself, the impact on any preceding production phases (for example, agricultural inputs) and the impact associated with the end-of-life of the wasted food. In the entire life cycle of a food product, the production phase has the most significant impact on natural resources. However, each phase carries additional environmental impacts. This means that the further a product progresses along the supply chain before it is lost or wasted, the greater its environmental cost or impact. This suggests that the further one is down the supply chain (for instance, consumption), the larger the food wastage footprint (Reynolds et al, 2019).

The hotspots for food waste along the supply chain vary geographically. Depending on the country, food waste occurs at different stages of the supply chain. Indeed, in developing countries, food waste tends to occur higher upstream (agricultural production, post-harvest handling and storage) while in developed countries food waste primarily occurs during the production, processing, distribution and consumption phases. In low-income regions, food waste is mostly caused by financial constraints; that is when producers are unable to purchase inputs or face structural limitations that affect harvest techniques, storage facilities, infrastructure, cooling chains, packaging and marketing systems. These limitations, coupled with climatic conditions conducive to food spoilage, lead to large amounts of food losses. In middle and high-income regions, food waste is caused by wasteful practices in the food industry and by consumers (both households and catering services). The food industry maintains strict retail cosmetic standards related to size and appearance and can cancel forecast orders, while insufficient purchase planning, as well as confusion over expiration date labelling, fosters high food waste. Understanding the different factors that facilitate food waste is important to better target food wastage reduction strategies (Herzber, Trebbin and Schneider, 2023).

The production of some products consumes more natural resources than others. Not all commodities are wasted in the same amounts, nor do they require the same number of natural resources to be produced. For instance, growing barley (500 grams), which requires 650 litres of water, is much less water-intensive than producing beefsteak (300 grams), which requires over 4,000 litres of water (waterfootprint network, .n.d.).

To effectively tackle food waste, it is important to understand where the waste hotspots are, both along the value chain and geographically, as well as which types of food commodity waste have the greatest impact in terms of natural resources. Figure 7.2 shows how The causes of food waste can be identified across four main different sectors:

1 production/storage
2 processing/transportation
3 wholesale and retail/food services
4 household consumer/end of life

In the production/storage stage, food waste can arise from overproduction due to the pressure to fulfil contractual obligations, strict appearance quality standards for produce, damaged products, inexpensive disposal alternatives and inedible parts of produce. In the processing/transportation stage, factors such as temperature fluctuations leading to spoilage, high aesthetic standards expected by consumers and retailers, packaging defects rendering produce unsaleable, oversupply due to consumer choices and overstocking due to poor planning and excess surplus contribute to food waste.

Figure 7.2   Food waste along the supply chain

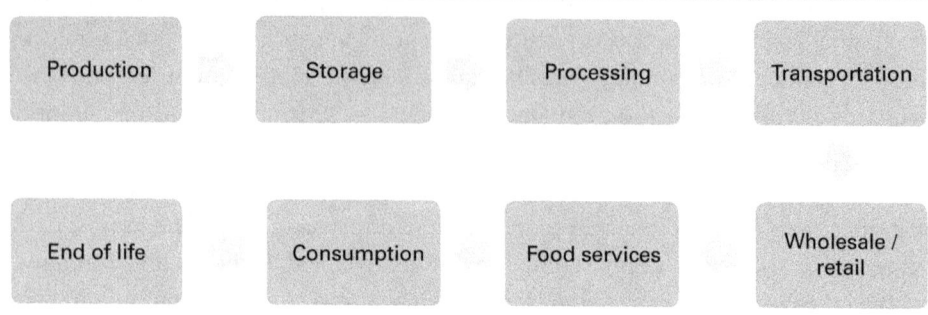

Adapted from Harvey, 1982

In wholesale and retail/food services, the lack of flexibility in portion sizes, inadequate planning in forecasting and ordering ingredients, consumer attitudes towards taking leftovers home, and refusal of food due to not meeting customer preferences can lead to significant food waste. At the household consumer/end of life level, food waste can be attributed to: buying in excess due to poor planning, improper storage resulting from lack of awareness, confusion over freshness and safety labels, discarding edible parts of produce like bread crusts or apple peels, discarding leftovers and large portion sizes (Schanes, Dobernig and Gözet, 2018).

The food waste situation in the UK

The United Nations' (UN) Sustainable Development Goal (SDG) 12.3 sets a target to cut per capita food waste and reduce food loss by half by 2030. The UN has calculated that in 2019, households, retail establishments and the food service industry globally generated a total of 931 million tonnes of food waste (UNEP, 2024). However, the UN has noted that obtaining accurate global estimates of food waste is challenging due to the current low availability of global food waste data and the highly variable measurement approaches. The UK is recognized as a country with high-quality data on this issue.

In the context of the UK, the most recent report by the charity Waste and Resources Action Programme (WRAP, a UK based charity) underscores the importance of continued efforts in reducing food waste across all sectors. In its 2018 report, WRAP states that almost 10 million tonnes of food waste were produced, two-thirds of which were intended to be for human consumption, and the remaining third was categorized as inedible food. This food waste broken down by sector indicates that the majority of food waste comes from households, contributing

6.6 million tonnes, which is 70 per cent of the UK's total food waste. Regarding the manufacturing and services sector, this contributes 1.5 million tonnes (16 per cent) to the total food waste. Furthermore, the hospitality and food service sector add another 1.1 million tonnes (12 per cent). This could be due to portion sizes being too large, customers not finishing their meals or food preparation waste. Finally, the retail industry accounts for a smaller portion of the total, contributing 0.3 million tonnes (3 per cent). This could be due to products not being sold before their expiry date, damage during transportation and handling, or overstocking (WRAP, 2024).

The report highlights the significant economic value of this waste, amounting to over £19 billion annually. This represents a substantial loss of resources. In terms of environmental impact, food waste is associated with more than 25 million tonnes of GHG emissions, contributing to global warming and climate change. Additionally, it is important to note that a significant portion of this wasted food, 6.4 million tonnes, could have been eaten. This is equivalent to over 15 billion meals. This indicates a significant opportunity to reduce waste and feed more people. However, the report also provides some positive news, as there has been a 15 per cent reduction in total food waste between 2007 and 2018. This includes an 18 per cent reduction in household waste. If the current rates of progress are maintained, the UK is on track to meet the UN's SDG 12.3, which aims to halve per capita global food waste at the retail and consumer levels by 2030 (Xameerah et al, 2024).

This progress is also reflected in local industry leadership. As regional hospitality associations took stock of sustainability priorities during the pandemic, food waste reduction emerged as both a moral and operational imperative.

INDUSTRY VOICE: ADRIAN ELLIS

Adrian Ellis, director of Hospitality Connect and former chair, Manchester Hoteliers Association, says 'At the Manchester Hoteliers Association, sustainability is one of our four strategic pillars, alongside business, recruitment, and charity. During the pandemic, we doubled down on our efforts to reduce food waste and support greener hospitality operations. Tackling food waste isn't just about saving money, it's about setting an example and building resilience. Through collaboration across more than 50 member properties, and partnerships with schools, charities, and the council, we're working to embed sustainability in the day-to-day fabric of hospitality.'

Impact of Covid-19 on food waste in the UK

The Covid-19 pandemic was instrumental in creating an overall awareness in the public and private sphere of how climate change, sustainability and our survival are interconnected (Aldaco et al, 2020). Notwithstanding the tragedy of the loss of human lives, there were positive lessons in terms of more environmentally and economically responsible food waste management.

In February 2021, WRAP released a series of surveys that explored the attitudes and behaviours of UK adults towards food waste during the unprecedented times of the Covid-19 pandemic in 2020. These surveys were conducted four times throughout the year (April, May, September and November), each time gathering responses from over 4,000 participants. The cumulative findings from these surveys indicated a downward trend in self-reported food waste levels. In November 2019, the self-reported food waste level stood at 24.1 per cent. Fast forward to November 2020, and this figure had dropped to 18.7 per cent. Interestingly, the lowest reported food waste level, 13.7 per cent, was observed in April 2020, coinciding with the first national lockdown. During the lockdown in April 2020, a significant 79 per cent of survey participants reported adopting new behaviours to manage food more efficiently. These behaviours included:

- more pre-shopping planning, such as checking what is already in the fridge and cupboards, reported by 41 per cent of respondents
- better food management at home, such as keeping an eye on use-by dates, reported by 35 per cent of respondents
- making more use of leftovers, a new habit started by 30 per cent of respondents

The report highlighted that these new behaviours persisted in the months following the first lockdown, indicating a potential long-term shift in attitudes towards food waste (WRAP, 2024).

The food waste situation in the UK THE sector

Food plays a critical role within the THE industry. It has an important environmental, economic and social function with profound implications.

The global food system is a significant contributor to environmental issues and resource depletion. The problem of food loss and waste intensifies some of the most critical challenges planet Earth is facing today, such as climate change, biodiversity loss and food security. Addressing these issues is a necessary step towards transforming food systems for improving the sustainability Triple Bottom Line, as understood by the UN themes of people, planet and prosperity. This position allows THE industry to influence both what is produced and sold by farmers and what consumers choose to buy and eat. Tourism is an integral part of the middle stages of the food value chain (food service), influencing both what agricultural producers sell and what consumers purchase. There is ample research from academics, international organizations and national agencies emphasizing an overall concerning picture that links food waste to food served in THE industry, especially when considering that according to the UNWTO (2024), an estimated 80 billion meals were served to international and domestic tourists in 2019.

Confronting food waste provides tourism entities with the opportunity to boost their operational proficiency, reduce expenses and mitigate their ecological impact. This encompasses curtailing GHG emissions via the execution of food waste minimization tactics and consolidating food security through the reallocation of surplus food. To pivot the focus of national tourism strategies from a rivalry-centric approach associated with gastronomy towards a sustainable food handling approach, underscoring the reduction of food waste can provide businesses with opportunities to expand their market while meeting their sustainability agenda. The THE sector partakes in various facets of food handling: procurement, stock control, menu crafting and display, consumption by patrons and waste handling. Food waste is an omnipresent issue that takes place at each of these stages. The reduction of food waste is an essential element in advancing towards a more sustainable modus operandi in food handling within the tourism sector, which could include the incorporation of circular methodologies.

While one of the most exhaustive pieces of research about food waste in THE in the UK dates back to 2013 (WRAP, 2013), it still provides a snapshot of areas of improvement and areas of good practice within the hospitality and food service sector.

The financial burden of food waste was around £2.5 billion per year in 2011, escalating to £3.0 billion per year by 2016. The report summarizes and underscores the potential to reduce waste and conserve financial resources. Each year, food and beverage (F&B) outlets generate a total of 2.871 million tonnes of waste, including food, packaging and other 'non-food' waste. Nearly half of this (46 per cent) is either recycled, processed through anaerobic digestion (AD) (He et al, 2024) or composted. Interestingly, each year, 920,000 tonnes of food is discarded at outlets, three-quarters of which could have been consumed. Carbohydrates, including potatoes, bread, pasta and rice, make up 40 per cent of the food waste. Reducing carbohydrate waste would significantly decrease the overall volume of food waste. The food wasted each year in the UK equates to 1.3 billion meals or one-sixth of the 8 billion meals served annually.

On average, 21 per cent of food waste is due to spoilage, 45 per cent results from food preparation and 34 per cent comes from consumer plates. Only 12 per cent of all food waste is recycled. In addition to food waste, 1.3 million tonnes of packaging (for food and drink and other non-food items used within hospitality and food service) and 0.66 million tonnes of other 'non-food' wastes are also discarded. This includes items like disposable kitchen paper and newspapers. Of the packaging and other 'non-food' waste, 62 per cent is recycled. The highest recycling rates are observed for glass and cardboard. Interestingly, 56 per cent of the discarded packaging and other 'non-food' waste could have been easily recycled (RSM Consulting, 2023).

The current state of excessive food waste within the THE sector presents a disconcerting, economically and environmentally untenable scenario. The severity of this issue necessitates immediate attention and action. However, it is crucial to note that this narrative does not conclude here. The next section offers some pragmatic and innovative recommendations to improve this situation. These solutions should provide the reader with potential remedies, hope, and a roadmap towards a more sustainable future within the THE industry. The section below explores these strategies in more detail and ascertains how to reduce food waste significantly.

THE food waste

In a 2021 UNEP report on the value-chain approach, which included food, an approach to identify areas of high resource use and environmental impact is provided with key intervention points for more sustainable consumption and production. The report found that while most resource use and environmental impacts occur at the primary production stage, producers have limited influence over these systems. Interestingly, a smaller number of actors in other stages of the food value chain, including food companies and retail services, hold significant power and largely dictate what is produced, sold and ultimately consumed (UNEP, 2024).

Hence, the mitigation of food waste stemming from THE-related functions and actions falls within tourism entities' remit. These organizations have the direct ability to implement effective food waste solutions in their operations. On the other hand, the impact of tourism on reducing food loss is more nuanced and challenging to quantify. However, by adopting sustainable practices and promoting responsible consumption, the tourism industry can indirectly contribute to food loss reduction. This dual approach not only addresses the immediate issue of food waste in THE operations but also influences the broader food value chain towards sustainability. It is a testament to the potential of the THE sector to be a catalyst for change in the global fight against food waste (see Figure 7.3).

Food waste solutions

The process of waste prevention within the THE sector necessitates a comprehensive approach, encompassing everything from menu planning to consumption. Looking at food preparation within THE sector-wide average, 21 per cent of food waste is attributed to spoilage, 45 per cent to food preparation and 34 per cent to consumer plates (WRAP, 2013; WRAP; 2024). These proportions, however, are

Figure 7.3 THE in the food value chain

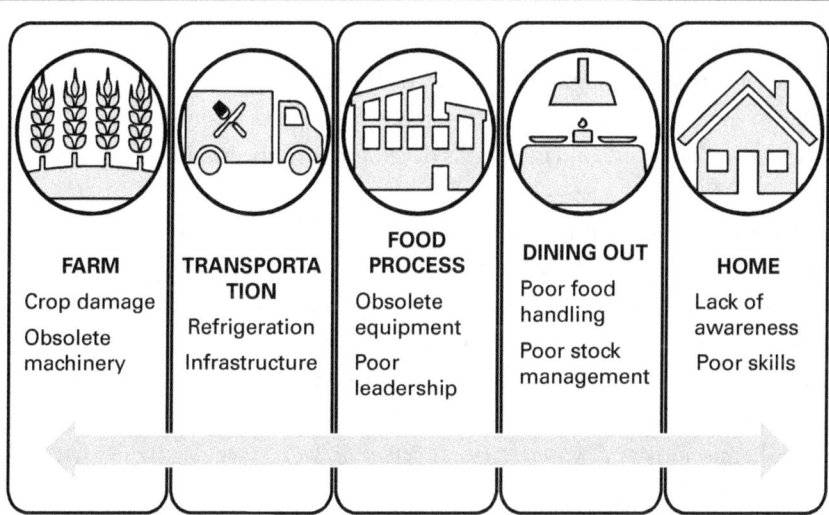

Adapted from UNEP, 2024

subject to variation depending on the specific kitchen operations and the extent of pre-prepared food usage. Various factors such as clients, suppliers, staff, budget and consumers significantly influence food waste generation. Food waste can potentially be produced at each critical stage of meal production. Therefore, an essential aspect of waste monitoring is to pinpoint where and why waste is produced. This understanding allows for the identification of priority areas that can be targeted to minimize waste generation. Moreover, other factors such as the kitchen layout, facilities and available space also play a role in waste generation. For instance, sites with limited space are more likely to depend on pre-prepared foods, which can contribute to waste. Thus, a holistic approach is required to effectively address food waste in the THE sector.

UK Food Waste Reduction Roadmap: The Guardians of Grub campaign

WRAP collaborates with the THE and food service sector, focusing on food, drink and plastics. The charity's mission envisions a sustainable world where resources are utilized efficiently. WRAP brings together companies from the THE and food service industries to work towards shared goals under initiatives such as the UK Food and Drink Pact (formerly known as the Courtauld Commitment – see Further resources) and the UK Plastics Pact.

Furthermore, the charity is actively involved with the sector through the UK Food Waste Reduction Roadmap. This initiative aids food businesses in taking specific actions to minimize waste within their operations, supply chains and consumer base, aligning with the UN's Sustainable Development Goal 12.3. Drawing from the UK Food Waste Reduction Roadmap in the Courtauld Commitment 2030, WRAP has launched the Guardians of Grub campaign. This initiative for reducing food waste in the sector is adaptable to organizations of all sizes and types. It supports businesses by providing free resources, including tools, guidance, skills development opportunities and campaign materials (WRAP, 2024).

The Courtauld Commitment 2030 (WRAP, 2024), now known as the UK Food and Drink Pact, is a voluntary agreement initiated by WRAP. The pact is designed for businesses across the entire UK food chain, including manufacturers, retailers, the THE industry and food service companies. The aim is to achieve sustainable use of resources and help the UK food and drink sector meet global environmental goals. The pact focuses on delivering farm-to-fork reductions in food waste, greenhouse gas (GHG) emissions and water stress. It sets ambitious targets aligned with UK and international environmental goals. The pact encourages collaborative action and innovation among businesses. It provides a framework for companies to take targeted action to reduce waste in their operations, supply chains and consumer base.

The current phase builds on the success of previous agreements since 2005. The targets are set for achievement by 2030. The pact is applicable across the UK, but its influence extends globally as it supports the UN's Sustainable Development Goal 12.3. In addition, WRAP's Guardians of Grub campaign is a key part of the commitment.

The Guardians of Grub Campaign, an initiative by WRAP, is designed for professionals across the THE in the food service sector. The campaign aims to tackle the significant issue of food waste, which costs the UK hospitality and food service sector billions each year. The campaign educates professionals to on how reduce the amount of food thrown away in their businesses. It provides free, practical tools and resources to help businesses reduce food waste in their kitchens and achieve greater financial savings. The campaign operates on the principles of 'Target-Measure-Act', providing free online training to implement this approach. It aims to change behaviour and relies on self-commitment. Businesses completing the Guardians of Grub training can post their success on social media using the campaign logo as free accreditation. While the campaign is based in the UK, its principles and resources can be applied globally to any establishment where food is served to order (WRAP, 2024).

To illustrate how these principles of targeting, measuring and acting on food waste can be effectively implemented in a large-scale tourism and hospitality setting, consideration a real-world example from P&O Cruises, facilitated by WRAP is useful. This example highlights the journey of a major cruise line in tackling food waste across its operations.

REAL-WORLD EXAMPLE   P&O Cruises: Navigating towards reduced food waste

Background

P&O Cruises, a cruise line operator, has embedded sustainability into its operational ethos, actively seeking solutions for positive long-term environmental impact. A key focus area has been the minimization of food waste across its fleet, without compromising the guest dining experience. Through daily monitoring and innovative strategies, P&O Cruises has made significant strides, demonstrating a strong commitment to food waste reduction.

Targets and commitment

P&O Cruises publicly committed to clear, progressive targets for reducing food waste per person, benchmarked against 2019 levels:

- By 2025: Achieve a 40 per cent reduction.
- By 2030: Achieve a 50 per cent reduction. As an interim achievement, the company successfully realized a 38 per cent food waste reduction per person.

The Target-Measure-Act approach in action

P&O Cruises adopted a data-driven approach, centred on the Target-Measure-Act framework, to guide its food waste reduction efforts:
Measure:

- **Daily weighing and monitoring**: A cornerstone of their strategy involved implementing a daily schedule for weighing food waste in the galleys (kitchens). This provided quantifiable data, enabling teams to identify areas needing attention.
- **Production vs. plate waste**: Both production waste (e.g. from food preparation) and plate waste (food left by guests) were weighed. This helped distinguish between kitchen inefficiencies (like knife skills, which weighing production waste could highlight) and issues with portion sizes (indicated by plate waste).
- **Tracking 'unusable' items**: Weighing items deemed 'unusable' provided valuable data to inform menu development and ingredient procurement, potentially leading to changes in specifications or finding uses for previously discarded parts.

Act – Improvements implemented: The data gathered through measurement informed a range of interventions:

- **Data-driven decisions**: A food waste dashboard was created, allowing to review focus areas and tailor interventions.

- **Targeted briefings**: Daily briefings with galley teams used the weighing data to discuss specific challenges and reinforce reduction goals.
- **Menu and service adjustments**: Insights from plate waste led to menu reviews and adjustments to dishes that were less popular. Procurement of smaller dishes for buffet areas also helped reduce waste.
- **Focus on problem items**: Specific items prone to wastage, such as butter, toast, jam and marmalade during breakfast service, received close monitoring, leading to further reductions.
- **Team engagement and awareness**: Emphasis was placed on fostering collaboration between front and back-of-house teams. Awareness campaigns 'Love Food, Hate Waste' were implemented for both crew and guests, encouraging mindful consumption. Some ships engaged in friendly competitions to reduce food waste more effectively.
- **Training and best practice**: Identifying areas like knife skills through waste analysis led to targeted training and sharing of best practices to improve efficiency.

Key learnings

P&O Cruises' experience underscores several critical lessons:

- Reliable data is fundamental to driving improvements and pinpointing areas for effective intervention.
- Collaboration across all teams (both front and back of house) is essential for success.
- Raising awareness and fostering a sense of collective responsibility among all team members is crucial.
- Visible monitoring, like daily weighing and displaying what has been wasted, can be a powerful tool for engaging teams and stimulating interest in reduction efforts.

This example demonstrates that through committed leadership, robust measurement systems, targeted actions and engaged teams, significant reductions in food waste are achievable even in complex operational environments like cruise ships.

REFLECTIVE QUESTION

Consider the strategies implemented by P&O Cruises in the real-world example above. Identify three specific actions taken by P&O that you believe could be effectively adapted and applied to reduce food waste in a different type of THE setting (e.g. a hotel, a contract caterer for events or a quick service restaurant). Briefly explain your reasoning for each.

In the next section are listed the principal interventions within the THE and food service sector and the potential avenues for waste reduction.

Food waste reduction interventions

It is important to note that a blend of strategies, tailored to the unique requirements of your specific organization, will likely yield the most effective results in waste minimization. This approach, endorsed by WRAP, emphasizes the importance of customization in waste reduction strategies. The economic viability of the various interventions outlined in this segment warrants meticulous scrutiny to ascertain their efficacy in achieving the anticipated outcomes. Monitoring and evaluation are integral to ensure that these eight interventions are not only cost-effective but also successful in reducing waste in line with set objectives.

Accurate forecasting intervention

Accurate forecasting intervention can help THE businesses strive to balance food production to avoid waste, yet the fear of underproduction, especially during peak times, can lead to overproduction. Quick service restaurants (QSRs) manage this by having high-demand items ready and using a 'make-to-order' system for specials and quieter periods. However, misjudged forecasts can result in waste. IT systems can predict demand based on historical sales data, aiding in supply ordering and portion control. Factors influencing demand include the size of the operation, menu, consumer numbers and preferences, weather and local events. Monitoring menu uptake and waste data can inform planning and reduce waste. Surplus food can be safely incorporated into the menu. The key is to minimize unpredictability as this helps with forecasting. Pre-selection of meals, or 'cook-to-order', can further reduce waste in predictable consumer bases, but timing is crucial.

Menu intervention

Menu intervention is crucial in minimizing food waste and is influenced by various operational decisions. Offering a diverse menu that satisfies consumers while reducing waste from less popular items is key. Mistakes can lead to unserved food being wasted, and menus may need to be refreshed regularly, especially in sectors where consumers frequent the same outlet. Contract caterers are often required to vary menus. There are nine main influences on menu planning, with some having more impact than others, such as budget. These are procurement, storage type and space, budget, service style and equipment, staff level, production equipment, production facilities, seasonality and consumer preferences (see Table 7.1 below).

Table 7.1 Nine influences on menu planning

#	Influences
1	Procurement strategies (e.g. supplier reliability, ingredient availability)
2	Storage type and available space (e.g. freezer, chilled, ambient capacity)
3	Budgetary constraints (e.g. cost per dish, overall food cost percentage)
4	Service style and equipment (e.g. buffet, à la carte, cook-chill systems)
5	Staffing levels and skills
6	Production equipment and its capabilities
7	Overall production facilities and layout
8	Seasonality of ingredients
9	Consumer preferences and dietary trends

Adapted from WRAP, 2024

Many THE food service outlets rotate their menus periodically, often seasonally, which can inadvertently cause waste as staff have fewer opportunities to optimize efficiency. Careful menu planning and review are vital in avoiding food waste. There is a 12-stage process used by THE professionals when designing a new menu, that by considering factors like service speed, consumer income, and seasonal food preferences can help reduce considerably food waste (see Table 7.2).

Procurement intervention

The procurement intervention of food and supplies significantly impacts waste generation in the THE sector. The temperature class of food products, whether chilled or frozen, affects storage duration and potential waste. Over-ordering due to inaccurate demand forecasting can lead to increased waste. While frozen ingredients offer flexibility, lack of freezer space may necessitate ordering chilled equivalents or in-house preparation. The proportion of expenditure on each temperature class varies across THE subsectors. QSRs, with their emphasis on pre-prepared and frozen ingredients, tend to generate less on-site waste but more upstream waste per meal than other subsectors.

The scale of procurement influences product pack sizes, delivery frequency and cost. The 'just in time' (JIT) delivery model allows THE food outlets to respond better to fluctuating consumer demand, but rising transport costs and carbon impacts may counterbalance this advantage. Packaging plays a crucial role in protecting and preserving food and drink throughout the supply chain. Optimizing packaging, involving suppliers and manufacturers, can reduce waste and carbon impact. Examples include glass bottle optimization, reviewing corrugate and carton board specifications, and

Table 7.2 12 stages of menu design

Stage	Activity	Intervention
1	Assess consumer group	Understand the preferences, dietary needs and expectations of the target consumer group.
2	Consider budget and resources	Evaluate the financial resources available and the cost of ingredients, labour and other resources needed.
3	Consider menu policies	Take into account any existing policies related to menu planning, such as nutritional guidelines or sustainability commitments.
4	Decide on menu structure	Determine the layout of the menu, including the number and types of dishes to be offered.
5	Write the menu	Draft the menu, including descriptions of each dish, ingredients and prices.
6	Undertake menu analysis – nutrition and cost	Analyse the nutritional content and cost-effectiveness of each dish on the menu.
7	Make any changes	Revise the menu based on the analysis, making adjustments to improve nutrition, cost-effectiveness or other factors.
8	Trial the menu	Test the menu in a real-world setting, gathering feedback from consumers and staff.
9	Review the trial	Evaluate the results of the trial, identifying any issues or areas for improvement.
10	Implement the menu	Roll out the final version of the menu in the establishment.
11	Monitor waste and satisfaction levels	Regularly check the amount of food waste generated and measure customer satisfaction levels.
12	Review the menu	Periodically review the menu, making adjustments as needed based on feedback, waste levels and other factors.

Adapted from WRAP, 2024

using reusable packaging. Bulk packaging formats can result in less packaging per unit of product but are only suitable for regularly used products at larger outlets with sufficient storage space.

Increasing the recycled content of packaging can also reduce its carbon impact. There are protocols available online for calculating the recycled content within the packaging (such as WRAP, 2024). Notably, green glass can have a higher recycled content than clear glass, lowering the carbon impact when sourced from the UK.

Meal intervention

Meal intervention significantly influences waste generation in the THE sector. Traditional kitchens prepare meals from scratch ('cook-serve'), leading to unavoidable waste like vegetable peelings. Many outlets now use pre-prepared components, often frozen for longer shelf-life. Both methods can coexist in a single outlet.

Pre-prepared items offer several advantages: they reduce preparation waste, have a longer shelf-life, require less energy and water if stored at ambient temperature, simplify cooking methods and allow greater flexibility in responding to fluctuating consumer demand. They also result in less waste due to more automated upstream manufacturing processes and lower transportation costs per meal. However, pre-prepared items also have disadvantages: they can involve more packaging per meal, be perceived as lower in quality and freshness, be more expensive, incur higher energy costs if stored chilled or frozen, offer less flexibility for reuse, may have minimum order quantities and can limit the use of the local supply chain.

Plate intervention

Plate intervention targets plate waste reduction. Over-serving can lead to waste, representing an avoidable cost. Operators often rely on their experience to determine portion sizes. Accurate portion control is challenging in self-service buffets, requiring a balance of choice and quality. Experience and new approaches can help reduce waste. In certain situations, like staff restaurants, reducing buffet items towards the end of service can minimize waste. However, this requires agreement between the client and the caterer and may not be feasible in all settings. Using smaller plates for buffets has been shown to reduce food waste.

Guest intervention

Guest intervention can significantly impact food waste at THE food outlets. The intervention can set up specific requirements with clients, especially within events and catering. These requirements, which may include quality, cost, food safety, nutritional value, choice range, service frequency, local sourcing, sustainability and approved suppliers, can help reduce waste. However, a mismatch between client stipulations and consumer preferences can contribute to waste. The growing preference for fresh produce, particularly chilled items with shorter shelf lives like salads, can lead to waste, especially in schools, prisons or large public institutions, such as hospitals, where healthier options are required. To reduce waste, it could be beneficial to identify popular choices among these healthier options and prioritize them.

Disposal intervention

Disposal intervention addresses waste disposal, which can be a significant cost. This cost can be mitigated by reducing waste at various stages: spoilage, preparation, unserved food and plate waste. Spoilage waste occurs when food exceeds its use-by date or is damaged. Preparation waste can reveal inefficient practices and cooking mistakes. Unserved food, edible but surplus to requirements, represents a significant financial loss. Plate waste arises when consumers leave food uneaten due to over-portioning or other food issues.

Sink disposal units (SDUs), while offering operational advantages, pose challenges to waste monitoring due to 'invisible' waste, subsequent lack of awareness, additional utility costs and regulatory compliance issues. Effective waste management requires a comprehensive approach. Reducing waste generation at the source, recycling, and composting are fundamental strategies. Implementing waste management systems can optimize these efforts. However, it is crucial that these systems are used correctly and consistently, and that all staff are trained and committed to minimizing waste. Regular monitoring and feedback are also essential to identify areas for improvement and measure progress. By adopting these solutions, THE food operations can significantly reduce their waste, leading to cost savings and a more sustainable operation.

Packaging intervention

Packaging intervention including other waste in the THE food sector varies by sub-sector. However, all sectors can improve waste diversion from mixed residual waste bins and SDUs. Redistributing surplus food to charities like Action Against Hunger or Feeding Britain (or others like UNICEF, Teaffund, Plan Zheroes and FareShare) is one strategy to prevent food waste, ensuring compliance with food safety regulations. Changes in legislation such as mandating recycling and banning landfill and sewer disposal of food waste is another. Opportunities for improving waste management include correct bin usage, waste data monitoring, recycling, staff engagement in recycling initiatives, collaboration with neighbouring businesses for waste collections and careful selection of waste management companies.

After recognizing the significant environmental and financial impacts of food waste in THE, the next section discusses strategies for influencing behaviours towards more effective waste management. These insights applied to the broader context of sustainability in THE can provide students, lecturers and professionals in these fields with a foundational understanding of a more sustainable THE industry.

> **INDUSTRY VOICE: ANDREA ZICK**
>
> Andrea Zick, UK-based food systems researcher and a Guardians of Grub Ambassador, champions the vital role of culinary knowledge in tackling food waste. She states, 'Chefs can influence a huge change in habits around menu design and circularity. Every trainee chef must be taught to measure and monitor food waste, because seeing what's wasted means you tackle it.' Andrea emphasizes that practical training, like that supported by Guardians of Grub, empowers the next generation of chefs to innovate with often-wasted ingredients, fostering a mindset of 'healthy plates and a healthy planet' from the start of their careers.

The human factor in food waste reduction intervention

The final section of this chapter focuses on consumer and staff behaviour. Both play a significant role in the volume and type of waste, particularly plate waste. Consumers often do not always feel responsible for the food they leave uneaten. Consumers look to operators for solutions, such as offering refills or different portion sizes, clear communication about what is included in a meal and the option to build their own meals. More on this will be discussed later in this chapter.

Staff behaviour plays a crucial role in waste management. Clear communication about waste reduction and recycling objectives enables staff to inform consumers and promote waste prevention. Training staff on waste generation areas and reduction opportunities, including the use of doggy bags/boxes, surplus food redistribution and correct portioning, can prevent inadvertent food wastage. Making staff aware of key areas of waste generation and opportunities to reduce it can encourage members of the team to help tackle waste. This includes promoting the use of doggy bags/boxes, awareness of surplus food redistribution and correct portioning.

A study by Chawla et al (2022) highlights that many staff members in the THE food outlets see waste prevention more as a money-saving strategy than a moral duty. This practical approach, driven by financial benefits, might not be as effective for waste prevention as intended. The study also points out that feeling a lack of control over behaviour could be a big hurdle, especially in the workplace. Employees might not always feel they can act in line with their attitudes due to company policies.

Social norms, or the unwritten rules of how to behave, can play a part in both creating and reducing food waste. These norms, which are part of the working culture in kitchens, can be strengthened by HR practices that recognize and reward individual achievements. The findings suggest that attitudes, social norms and perceived behavioural control all play a part in shaping intentions to act in an environmentally friendly way. In the workplace, social norms and perceived behavioural control might have a bigger influence on encouraging or restricting such behaviour. Hence, managers can play a key role in encouraging environmentally friendly behaviour by harnessing the power of social norms and fostering a sense of control among employees. The findings also suggest that environmentally friendly attitudes tend to cluster together, which could inform hiring practices in the hospitality industry. Waste reduction opportunities lie in providing guests with choices such as offering refills, varying portion sizes, and allowing (when and where possible) them to customize their meals. One strategy is to serve smaller portions initially, with the option for 'seconds', a practice adopted by ISS, a facilities management company and contract caterer in Denmark (Price, 2022). Encouraging consumers to take leftovers home in doggy bags or boxes is another effective strategy. Communication is key to successful waste reduction strategy with guests.

A study by Alsuwaidi et al (2022) explores key factors influencing guests' food waste reduction in the THE sector. It finds demographics, socio-economic factors, attitudes towards sustainability and anticipated feelings of guilt or pride significantly affect these behaviours. The study suggests that understanding guests' personal traits can help identify food waste behaviours that can help managers develop better strategies. Some examples are improving awareness campaigns and designing mechanisms to reduce food waste. Personal norms on waste reduction are a key predictor of guests' food waste reduction behaviour. While providing valuable insights, the study acknowledges limitations due to the complexity of accurately predicting human behaviours. Yet, the study recommends that THE food outlets develop marketing strategies to stimulate waste reduction as part of broader sustainability practices.

Messages should be brief and clear, considering that guests' attention might be fleeting. Using emotion can help earn engagement, capturing both the guests' hearts and minds. Positive instructions (like 'do', 'try' and 'help') are more impactful than negative ones (such as 'don't' and 'avoid'), as guests respond better to encouragement than shaming. Leveraging social influence can be beneficial, letting guests know they are part of a collective effort towards waste reduction. This can enhance their sense of accomplishment and even provide recognition by sharing their participation with others. Presenting compelling facts and straightforward data can be eye-opening for guests and lend credibility to the cause. Lastly, staff members should be engaged as ambassadors, equipped to answer guests' questions and offer encouragement. This multifaceted approach can significantly contribute to the goal of reducing food waste in THE food outlets.

Conclusion

Food waste is not just an issue of waste management, but it is intertwined with the climate crisis, responsible and sustainable consumption, and the sustainability of the THE industry. It is a complex problem that requires a multifaceted approach, involving everyone from households to supply chains, from consumers to policymakers. The chapter encourages THE students, lecturers and professionals to take these insights forward, to innovate and implement effective strategies for waste minimization and sustainable consumption. The fight against food waste is a collective responsibility, and every action counts towards a more sustainable future.

> **ACTIVITY 1: DESIGN A FOOD WASTE REDUCTION PLAN FOR A THE TRADE**
>
> This activity invites you to apply your understanding from this chapter to analyze common food waste challenges in the THE sector and propose practical solutions. It will help you apply the key concepts and food waste reduction strategies discussed in this chapter to realistic scenarios encountered in the THE sector. You will practice identifying potential causes of food waste and suggesting appropriate interventions discussed in the chapter.
>
> *Instructions*
>
> Read each of the three scenarios below. For each scenario, provide concise answers to the questions that follow, drawing upon the knowledge gained from Chapter 7 (including different types of interventions, the importance of measurement and human factors).
>
> 1. The hotel breakfast buffet
>
> A 150-room hotel offers a buffet breakfast. Management has noticed significant amounts of untouched pastries and bread rolls being discarded daily after the breakfast service, along with nearly full single-serving jam and butter portions.
>   Questions:
>
> 1 **Identify two likely reasons** (drawing from concepts in Chapter 7) for this specific type of food waste at the breakfast buffet.
>
> 2 **Suggest two practical interventions** (drawing from concepts in Chapter 7 or inspired by the P&O real-world example) that the hotel could implement to reduce this waste. Briefly explain your choices.

2. The independent café's lunch rush

'The Zest', a busy independent café, prepares a large batch of various sandwiches and salads each morning for the anticipated lunchtime rush. While sales are generally good, there are consistently 10-15 items left over at the end of the day that must be discarded due to freshness concerns.
Questions:

1 Thinking about the Target-Measure-Act framework, **which part of this process** could most help the café address its leftover food? Explain briefly.

2 **Propose two different strategies or interventions** the café could adopt to minimize the number of unsold sandwiches and salads being wasted.

3. The university event catering

A university's in-house catering team provided a hot buffet lunch for a departmental conference of 80 expected attendees. Final attendance was only 65, and a considerable amount of food, particularly the vegetarian main option and side salads, was left over in the serving dishes. Staff also observed many half-eaten portions of the meat main course on cleared plates.
Questions:

1 **Identify one issue** related to event planning/forecasting and **one issue** related to menu planning/consumer preference that likely contributed to the food waste.

2 **Suggest one strategy** to improve communication with event organizers for more accurate numbers in the future, and **one menu-related adjustment** the catering team could consider reducing plate waste at similar events.

Submit your answers to the tasks for all three scenarios, aiming for thoughtful but concise responses (approximately 100-200 words per scenario) that focus on applying the chapter's content; these responses can subsequently be peer assessed.

KEY TAKEAWAYS

- Significant global impact:
  - Food waste is a critical global challenge with severe environmental (significant carbon footprint, resource degradation), economic and social (e.g. increased food insecurity) consequences.

- Understanding the terminology:
  - Differentiating 'food loss' (mainly in production/post-harvest, common in developing countries) from 'food waste' (retail/consumption stages, higher in developed countries) is vital for targeted supply chain solutions.
- The THE sector's crucial role:
  - The THE sector significantly contributes to food waste but also has a unique responsibility and opportunity to implement impactful reduction strategies.
- A systematic approach is key:
  - A structured, data-driven approach like 'Target-Measure-Act' can be key for THE businesses to understand waste, implement interventions and track progress effectively.
- Diverse practical interventions:
  - Numerous practical interventions, including strategic menu design, better forecasting, optimized procurement and refined preparation can prevent food waste in THE operations.
- The human factor is pivotal:
  - Addressing the human factor by engaging and training staff, and influencing consumer behaviour towards mindful consumption, is crucial for successful food waste reduction.
- Managing unavoidable waste sustainably:
  - Sustainable options like composting, anaerobic digestion or food donation are crucial for managing unavoidable food waste, diverting it from landfills and recovering value.

REFLECTIVE QUESTIONS

1 Across diverse THE operations, what is the most significant universal barrier to effective food waste reduction, and how can it be dismantled?

2 Reflecting on the 'human factor', what single behavioural change or leadership action in your future THE role or as a consumer would most positively impact food waste reduction, and why?

3 Beyond established interventions, what innovative technology or societal shift might revolutionize THE sector's food waste management in the next decade?

4 How can THE businesses extend their influence beyond operations to encourage more responsible food consumption habits among guests and the wider community?

Further resources

Blondin, S and Attwood, S (2022) Making food waste socially unacceptable: What behavioural science tells us about shifting social norms to reduce household food waste, World Resources Institute, 6 June, www.wri.org/research/making-food-waste-socially-unacceptable (archived at https://perma.cc/MF8E-QFKT)

WRAP (2022) The Courtauld Commitment 2030: Progress and Insights Report 2022, www.wrap.ngo/resources/report/courtauld-commitment-2030-progress-and-insights-report-2022 (archived at https://perma.cc/BF53-HAMC)

WRAP (.n.d.) HISTORY From Commitment to Transformation: The Evolution of the UK Food and Drink Pact, www.wrap.ngo/take-action/uk-food-drink-pact/history (archived at https://perma.cc/MRV4-SQJE)

References

Aldaco, R, Hoehn, D, Laso, J, Margallo, M, Ruiz-Salmón, J, Cristobal, J, Kahhat, R, Villanueva-Rey, P, Bala, A, Batlle-Bayer, L, Fullana-i-Palmer, P, Irabien, A and Vazquez-Rowe, I (2020) Food waste management during the COVID-19 outbreak: A holistic climate, economic and nutritional approach, *The Science of the Total Environment*, 742, p 140524, https://doi.org/10.1016/j.scitotenv.2020.140524 (archived at https://perma.cc/7BV5-JYNN)

Alsuwaidi, M, Eid, R and Agag, G (2022) Tackling the complexity of guests' food waste reduction behaviour in the hospitality industry, *Tourism Management Perspectives*, 42, p 100963

Amicarelli, V, Lagioia, G and Bux, C (2021) Global warming potential of food waste through the life cycle assessment: An analytical review, *Environmental Impact Assessment Review*, 91, p 106677, https://doi.org/10.1016/j.eiar.2021.106677 (archived at https://perma.cc/J2Q7-EAV6)

Chawla, G (2015) Sustainability in hospitality education: A content analysis of the curriculum of British universities. In F Bezzina and V Cassar (eds.) *Proceedings of the 14th European Conference on Research Methodology for Business and Management Studies*, University of Malta, Valletta, Academic Conferences and Publishing International

Eurostat (2023) Food waste per capita in the EU remained stable in 2021, 29 September, https://ec.europa.eu/eurostat/web/products-eurostat-news/w/ddn-20230929-2 (archived at https://perma.cc/QKC9-7TJZ)

FAO (2022) Statistical Yearbook World Food and Agriculture 2022, https://openknowledge.fao.org/handle/20.500.14283/cc2211en (archived at https://perma.cc/LGP4-WMG2)

Harvey, D (1982) The limits to capital. In P Hubbard, R Kitchin, G Valentine and N Castree (eds.) *Key Texts in Human Geography*, SAGE Publications, London, https://sk.sagepub.com/dict/edvol/key-texts-in-human-geography/chpt/limits-capital-1982-david-harvey (archived at https://perma.cc/SV6E-96JQ)

He, K, Liu, Y, Tian, L, He, W and Cheng, Q (2024) Review in anaerobic digestion of food waste, *Heliyon*, **10** (7), e28200, https://doi.org/10.1016/j.heliyon.2024.e28200 (archived at https://perma.cc/9G6A-LKHS)

Herzberg, R, Trebbin, A and Schneider F (2023) Product specifications and business practices as food loss drivers – A case study of a retailer's upstream fruit and vegetable supply chains, *Journal of Cleaner Production*, **417**, p 137940, https://doi.org/10.1016/j.jclepro.2023.137940 (archived at https://perma.cc/J5LU-WMYJ)

IPCC (2022) Global warming of 1.5°C: IPCC special report on impacts of global warming of 1.5°c above pre-industrial levels in context of strengthening response to climate change, sustainable development, and efforts to eradicate poverty, Hanoi University Of Science And Technology, http://dlib.hust.edu.vn/handle/HUST/21737 (archived at https://perma.cc/7X8Q-YYET)

Price, K (2022) ISS commits to 50% reduction in food waste by 2027, The Caterer, 9 February, www.thecaterer.com/news/iss-commits-50-percent-reduction-food-waste-2027-cool-pledge (archived at https://perma.cc/QT9F-XTS6)

Reynolds, C, Goucher, L, Quested, T, Bromley, S, Gillick, S, Wells, V K, Evans, D, Koh, L, Kanyama, A C, Katzeff, C, Svenfelt, A and Jackson, P (2019) Review: Consumption-stage food waste reduction interventions – What works and how to design better interventions, *Food Policy*, **83**, pp 7–27, https://doi.org/10.1016/j.foodpol.2019.01.009 (archived at https://perma.cc/4YJK-CA5J)

RSM Consulting (2023) Alternatives to single-use plastics in food packaging and production, Food Standards Agency, 31 August, https://doi.org/10.46756/sci.fsa.taf512 (archived at https://perma.cc/K9EK-V2JY)

Schanes, K, Dobernig, K and Gözet, B (2018) Food waste matters – A systematic review of household food waste practices and their policy implications, *Journal of Cleaner Production*, **182**, pp 978–91, https://doi.org/10.1016/j.jclepro.2018.02.030 (archived at https://perma.cc/6NYM-73DZ)

Symeon, G K, Akamati, K, Dotas, V, Karatosidi, D, Bizelis, I and Laliotis, G P (2025) Manure management as a potential mitigation tool to eliminate greenhouse gas emissions in livestock systems, *Sustainability*, **17** (2), p 586, https://doi.org/10.3390/su17020586 (archived at https://perma.cc/YUG9-KS9Y)

UNEP (2016) The State of Biodiversity in Latin America and The Caribbean: A mid-term review of progress towards the Aichi Biodiversity Targets, Convention on Biological Diversity, www.cbd.int/gbo/gbo4/outlook-grulac-en.pdf (archived at https://perma.cc/4S9N-BKYG)

UNEP (2024) Food Waste Index Report 2024. Think Eat Save: Tracking Progress to Halve Global Food Waste, March, https://wedocs.unep.org/xmlui/handle/20.500.11822/45230 (archived at https://perma.cc/7D4Z-64AC)

UNWTO (2024) Statistical Framework for Measuring the Sustainability of Tourism (SF-MST): Final draft prepared for UN Statistical Commission, www.unwto.org/tourism-statistics/statistical-framework-for-measuring-the-sustainability-of-tourism (archived at https://perma.cc/N84X-XZFN)

Waterfootprint network (.n.d.) WATER Virtual Water Embedded in Product FOOTPRINT, www.waterfootprint.org/resources/schoolresources/Poster-A3-WaterFootprint-of-Products.pdf (archived at https://perma.cc/GW2W-AZVS)

WRAP (2013) Where food waste arises within the UK hospitality and food service sector: Spoilage, preparation and plate waste, www.wrap.ngo/sites/default/files/2021-03/Where-food-waste-arises-within-%20hospitality-and-food-service.pdf (archived at https://perma.cc/TYE6-HPBW)

WRAP (2024) Courtauld Commitment 2030 Annual Report 2023-2024, Autumn, www.wrap.ngo/sites/default/files/2024-12/241205-WRAP-Courtauld-2030-Annual-Progress-Report.pdf (archived at https://perma.cc/K8UJ-EGXS)

Xameerah, M, Smith, L, Stewart, I and Burnett, N (2024) Food waste in the UK: Research briefing, 12 April, https://researchbriefings.files.parliament.uk/documents/CBP-7552/CBP-7552.pdf (archived at https://perma.cc/A3KT-52FX)

Yang, Y, Liu, X, Chen, Y, Xu, Q, Dai, Q, Wei, H, Xu, K and Zhang, H (2024) Environmental impact assessment of rice–wheat rotation considering annual nitrogen application rate, *Agronomy*, **14** (1), p 151, https://doi.org/10.3390/agronomy14010151 (archived at https://perma.cc/7CZB-3FAW)

8 | Basics of sustainable menus: Principles and practices

 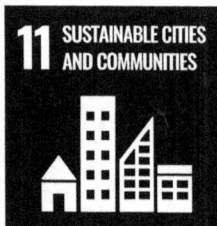  

| Systems thinking competency | Critical thinking competency | Strategic competency | Normative competency |

CHAPTER AIM

This chapter aims to develop the thinking needed when developing a sustainable menu. It provides a practical understanding of the key factors involved in the crucial preliminary decision-making process when planning, designing and preparing a sustainable menu. This chapter aims to develop the thinking needed when developing a sustainable menu within the tourism, hospitality and events (THE) industry. It provides a practical understanding of the key factors involved in the crucial preliminary decision-making process when planning, designing, and preparing sustainable food offerings. Emphasis is placed on aligning menu choices with key global frameworks, including the Sustainable Development Goals (SDGs), particularly SDG 12 (Responsible Consumption and Production), SDG 13 (Climate Action), SDG 2 (Zero Hunger), and SDG 11 (Sustainable Cities and Communities), addressing the significant impact of food systems.

The chapter guides learners in applying key sustainability competencies such as systems thinking, strategic, critical thinking and normative competencies to menu development, encouraging ethical sourcing, waste reduction and the promotion of healthier, planet-friendly diets.

LEARNING OUTCOMES

Upon completion of this chapter, you will be able to:

- understand the concept of sustainable menus
- identify the basics of menu planning, design and preparation
- recognize the operational factors that impact the development of a sustainable menu
- critically evaluate the ethical considerations surrounding sourcing when developing a sustainable menu

KEY WORDS

Ethical sourcing, food miles, menu layout, planetary-healthy diet, sustainable menu, three-step approach.

Introduction

This chapter provides instrumental knowledge for integrating sustainability into menu development. It begins by exploring the strategic and communicative functions of a menu before delving into key considerations for sustainability, including ingredient sourcing strategies (local, seasonal, organic, fair trade), menu layout techniques to encourage sustainable choices, and the concept of food miles. Furthermore, the chapter introduces the Planetary Health Diet as a framework for balancing human and environmental health, outlines a practical three-step approach (Planning, Designing, Presenting) for sustainable menu creation, and discusses the importance of aligned sustainable operations and building local networks. An activity is also included to reinforce understanding of sustainable menu principles.

The menu plays a central role in promoting sustainability within the food and beverage industry, yet its impact is often overlooked. As one of the primary marketing tools of any food establishment, a menu directly influences every customer's choice. However, to provide a comprehensive guide to creating a menu from start to finish is beyond the scope of a single chapter. Instead, this chapter introduces key sustainability principles to integrate into the three main stages of menu development: planning, designing and presenting. Before exploring these principles, it is important to have a clear understanding of what is meant by a sustainable menu and why it is an integral part of the food and beverage operations in the THE industry. Huang, Hall and Chen (2023) define a sustainable menu as one designed to promote environmentally friendly and sustainable food choices. In the THE industry, where competition is fierce and trends shift rapidly, businesses may prioritize achieving the bottom line; in other words, achieving 'profit' at all costs, over sustainability. However, incorporating sustainability into menus not only aligns with cost-saving strategies but also attracts environmentally conscious customers.

By adopting a long-term sustainable approach, food and beverage businesses can appeal to new customers while reinforcing their commitment to sustainability for existing patrons. Professionals widely agree that such menus are crucial in guiding and influencing consumers' decisions in full-service restaurants (Huang, Hall and Chen, 2023). These menus can be more cost-effective, scalable and influential in promoting sustainable diets, and encourage personal behavioural changes (Attwood, Chesworth and Parkin, 2020).

This chapter will explore strategies for developing a more sustainable food and beverage offering through effective menu planning, design and presentation, demonstrating that the principles can be applied to most menu types. The next section explains why the menu matters before we turn to how to make it sustainable.

The role of the menu

With the world's population projected to grow from 7 billion to an anticipated 9.8 billion by 2050, we are on track to see a surge in overall food demand by more than half, and a nearly 70 per cent increase in demand for foods derived from animals (Rockström et al, 2023). Food production is responsible for a quarter of annual greenhouse gas emissions (Rockström et al, 2023). This scenario underscores the pressing need for sustainable practices in our food systems.

The challenges of increasing food demand and environmental concerns demonstrate how sustainable menus can serve as a key strategy in the transition towards more responsible consumption and production in the food and beverage industry (Lera, 2019). The role of sustainable menus in this context becomes crucial. Menus are not just a list of offerings at a food establishment, but a powerful tool that can influence consumer choices towards more sustainable consumption.

The menu is the crucial first point of contact, where potential sales are either made or missed. If customers are attracted to the offerings and find the prices reasonable, they are likely to enter the establishment with the intention to make a purchase. While the quality and delivery of the menu items are important, the menu should reflect and align with the sustainability ethos of the organization. Therefore, the menu has a strategic function as well as a communicative one.

The strategic function of a menu

The menu sets a clear objective that steers the efforts of all the business practices. For clarity and consistency purposes, the professional charged with thinking strategically about menu planning (whether head chef, restaurant manager, food and beverage manager or business owner) has been named henceforth 'menu planner'. The menu planner will have to follow several stages when thinking strategically about menu development to ensure the best quality product at the most competitive price. The recommended stages are:

1 Identify the target market and market segmentation to understand the target market's food and beverage expectations.
2 Research the ingredients for the menu items, making sure that they are easily obtainable and cost-effective.
3 Determine that the menu items can be prepared and presented to appeal to the target market.
4 Assess whether the menu effectively meets the needs of the target market.
5 Check that the menu aligns with the operation's goals while attaining its sustainability ethos (Ip and Chark, 2023).
6 Use the menu as a strategic tool to help achieve the operation's objectives.

This is especially important in the context of sustainability, where every decision can have significant impacts.

The communicative function of a menu

The menu layout informs guests about the establishment's offerings, including dishes, beverages and unique items. The menu plays a crucial role in enticing customers to make purchases through appealing descriptions and strategic item placement. Most importantly, the menu communicates the operation's commitment to sustainability, reflecting its identity, values and contribution to environmental responsibility. For instance, many establishments may serve sustainability-oriented food items, but a sustainability-oriented menu planner owner will use the menu to detail how their food (and beverage) items are sustainably sourced and prepared. They can highlight the green credentials of the supplier, the lower carbon footprint of ingredients or spices used, the lower energy use of cooking methods, or the variety of local condiments offered. All these details can be shown on the menu.

Similarly, the choice of menu items can showcase the operation's support for local produce, fair trade and ethical sourcing. Moreover, the menu can serve social purposes; for instance, it can inform guests about health considerations, such as calorie count or risks associated with consuming certain food, such as red meat (Wansink and Love, 2014).

Using recycled or sustainably sourced paper for the menu, or opting for digital menus to reduce paper waste, can reinforce the operation's sustainability ethos. Attention to detail and a true commitment to sustainability are crucial here. Chapter 9 details how conflicting messages can erode trust and significantly undermine sustainability efforts. For instance, efforts to reduce carbon footprint should not be limited to tracking the food miles of suppliers. It should encompass all menu components, including paper, ink and other materials, not just food and beverages. (For more information about reducing carbon footprint, please refer to Chapter 5).

Considerations for a sustainable menu

When planning, designing and preparing a menu, a number of key considerations have to be addressed if the menu is to function as the business intends. The sourcing of ingredients, seasonality of produce and ethical certification and trading practices of suppliers have to form part of the decision making process for a sustainable menu.

Sourcing ingredients

Sourcing sustainable ingredients for a menu is essential for environmentally friendly dining practices. However, the integration of sustainable options into food and beverage menus is often perceived as a costly venture. While sourcing local, organic and ethically traded products can initially present financial and logistical challenges, it is important to note that not all sustainable choices are more expensive. Some can even be cost-effective. Strategic modifications can be made to balance costs across the menu, allowing for investment in sustainable options (Attwood, Chesworth and Parkin, 2020). For instance, reducing waste and targeting expensive products or processes for cost reduction can provide flexibility in selecting sustainable food items.

Sustainable sourcing means meeting the business' needs for goods and services in a manner that is environmentally friendly, economically viable and socially beneficial. This approach ensures that operations are not only cost-effective but also contribute positively to the broader community, balancing environmental stewardship, financial viability and societal value within an overarching sustainability strategy.

It is important to note that many consumers are becoming more conscious of the environmental impact of their food choices and are actively seeking out establishments that prioritize sustainability. By emphasizing which ingredients are ethically sourced and how they are sourced, businesses can appeal to these environmentally aware customers and differentiate themselves in the market.

As Table 8.1 illustrates, the breadth of sustainable ingredient options for a sustainable menu is useful for illustrating these benefits.

The importance of locally sourcing menu ingredients

Locally sourcing menu ingredients is a key component of sustainable menu practices offering environmental, economic and social (community) benefits. By prioritizing local suppliers, food and beverage establishments can reduce their environmental impact, support regional economies and strengthen community links. This approach not only promotes sustainability in menu planning, design and preparation but also develops strong relationships between business and local producers to benefit the environment along the supply chain (Filimonau and Krivcova, 2017).

Environmental benefit

Locally sourcing ingredients plays a significant role in reducing the environmental impact of food production and transportation. By sourcing ingredients locally, menu planners can work directly with farmers and producers to customize their orders to match their needs. By supporting local farms that use sustainable practices such as organic farming and regenerative agriculture that aim to mimic natural ecosystems through land management and conservation practices, food and beverage operators are engaging with sustainable practices at source.

When ingredients are sourced locally, they travel shorter distances from farm to table, greatly reducing fuel consumption and greenhouse gas emissions associated with long-distance transportation. By minimizing food miles – the distance food travels from its source to its destination – food and beverage establishments can lower their carbon footprint. Imported ingredients often travel thousands of miles, contributing to high levels of carbon emissions (food miles are discussed later in this chapter). In contrast, locally sourced ingredients have shorter supply chains, leading to lower emissions. An associated environmental benefit of local sourcing and transportation is the reduction in the energy required for food preservation (Kim and Hall, 2020).

Table 8.1 The environmental benefits of sustainable menu items

Sustainable options	Environmental benefit	Menu example
Locally sourced ingredients	Reduction in the carbon footprint associated with transportation.	A green salad made with locally grown vegetables.
Organic ingredients	• Grown without synthetic pesticides or fertilizers. • Support sustainable agriculture – soil health, biodiversity and water conservation. • Healthier to eat.	Organic quinoa and vegetable bowl.
Seasonal ingredients	Reduction for the need for long-term storage and transportation.	Pumpkin soup in autumn.
Plant-based options	Significantly reduce environmental impact compared to meat-based meals.	A lentil and vegetable curry.
Certified ingredients	Sustainability certifications, such as Fairtrade, Rainforest Alliance, or Organic JAS, ensure that the food is produced in an environmentally and socially responsible manner.	A dessert made with Fairtrade chocolate.
Sustainable seafood	Serving fish and shellfish from sustainable fishing or aquaculture practices helps protect marine ecosystems.	MSC-certified wild-caught salmon served with locally sourced organic potatoes.
Minimizing use of processed ingredients	• Replacing processed foods with fresh-cut vegetables, whole grains, beans and pulses is healthier. • Lower environmental impact as processing food requires extensive energy for production, packaging and preservation.	A wholegrain pasta with fresh tomato sauce.
Reducing food waste	Much of what is thrown away as food waste is edible. Creating dishes that utilize parts of ingredients that are often discarded, like vegetable stems and leaves, can help reduce food waste.	A pesto made from carrot tops.

Local produce is fresher and more likely to be delivered shortly after harvest, thus it requires less storage and refrigeration in comparison to ingredients sourced outside the area. Local produce can be delivered fresh without the need for prolonged preservation, reducing the need for refrigeration or freezing typically required for nationally transported or imported goods.

Historically, the THE sector is one of the biggest contributors to food waste (UKHospitality, 2022) through the discarding of surplus and imperfect food. Globally, food surplus and food imperfections have resulted in food waste becoming an environmental issue. On a local level, food surplus and imperfections refer to the excess or non-standard produce local farmers may have grown which can be used instead of being wasted (Imperfect Foods, .n.d.). By choosing to purchase surplus food, menu planners can reduce waste while benefiting from fresh, in-season produce.

Imperfect produce refers to perfectly edible fruits, vegetables and other agricultural products that do not meet standard aesthetic or size expectations for commercial sale to large suppliers such as supermarkets. Farmers often struggle to sell imperfect produce because many markets prioritize uniformity in appearance. However, the advantage for local food and beverage businesses is that they can purchase these items at lower cost. Using imperfect produce helps reduce food waste and allows farmers to profit from produce that may otherwise be discarded (Imperfect Foods, 2024).

In both cases, food surplus and imperfections can represent an opportunity for menu planners to provide a menu that supports sustainable practices, reduces food waste, lowers costs and supports local farmers to optimize their yields. (For more information about food waste, please refer to Chapter 7).

Economic benefit

From the perspective of the local economy, buying locally sourced ingredients directly benefits both the food and beverage business and the local community. Menu planners may find that purchasing ingredients from local farmers, producers and artisans is more cost-effective as the ingredients do not require the same level of transportation, storage or packaging. Without the need for long-distance shipping or extended refrigeration, operational expenses are lowered, making local sourcing an attractive option for menu planning.

Sourcing locally also shortens the supply chain, allowing businesses to build strong, direct relationships with suppliers. These relationships often lead to more flexible negotiations, enabling businesses to secure better prices and adapt to changes in supply. For example, during periods of scarcity or surplus, local farmers and menu planners can work together to find alternatives that meet both cost and quality requirements to the benefit of both parties.

By keeping the economic exchange within the locality also contributes to regional economic resilience, particularly in areas where agriculture is the primary industry. By supporting local producers and suppliers, food and beverage establishments are helping to sustain local jobs and the local community. Without that contribution, a restaurant or hotelier may find their operational practices undermined in times of economic downturn.

A cautionary note

While local sourcing is the preferred approach to acquiring ingredients for a menu, local sourcing is significantly influenced by price and availability. While menu planners may find local food to be of higher quality than imported options, the extent of local produce usage and quality varies region by region. In regions with limited agricultural production businesses may find local ingredients more expensive. Therefore, while it is possible that some managers can rely on organic farms to supply fresh produce to their establishments, others, such as cruise ships, may find it difficult to source sufficient quantities of local food from docking destinations, as small-scale producers often cannot meet the demand of large operations (Hoarau-Heemstra et al, 2023).

Tourism businesses, especially those in remote holiday destinations, face particular challenges in accessing local produce and establishing a reliable supply. The definition of 'local food' also varies depending on the location, with differing perspectives on what constitutes local. Defining 'local food' can be challenging because there isn't one single definition. Some approaches focus on distance, using 'food miles' as a guide – perhaps defining local as food travelling less than 100 miles. Another view centres on the supply chain, emphasizing a short, direct link between the producer and the consumer. Alternatively, 'local' might simply refer to food sourced within a certain geographical area, such as a specific region or even the whole country, depending on its scale (Smith and MacKinnon, 2009; Feldmann and Hamm, 2015; Schmitt et al, 2017).For smaller, more isolated locations, such as island hotels, the local radius for sourcing ingredients is typically much smaller than for land-based establishments in more densely populated areas.

To mitigate these challenges, the nearest food supply should be prioritized wherever feasible. Businesses can explore alternatives such as establishing or joining local food cooperatives, organizing local food festivals and learning from best practice hotels (Brune et al, 2023). Many food and beverage establishments set up their own kitchen gardens. Despite their small scale, these gardens offer numerous benefits including minimal transportation costs, reduced packaging, fresher produce and a better CO_2 balance. Additionally, food can be cultivated to suit the business's needs. A well-planned kitchen garden or even small-scale farm can also serve as an attraction for guests, enhancing their awareness of local sourcing and food cultures, and potentially increasing their willingness to pay for the food (Feldmann and Hamm, 2015).

Seasonal produce

Opting for seasonal food significantly reduces food miles and CO_2 emissions by eliminating the need for environmentally harmful transport and energy-intensive greenhouses. Seasonal fruits and vegetables, harvested at full maturity, are often healthier and tastier, supporting local producers and promoting local food culture (Stein and Santini, 2022).

Buffet offerings vary among businesses. While some, like cruise ships and certain hotels, maintain a standard buffet year-round, others emphasize seasonal fruits. The main challenges to incorporating a wider range of seasonal produce are higher costs and guest preferences.

Food and beverage businesses can adapt their offerings based on the season and creatively market their seasonal foods. For example, a mussels festival, which introduces guests to seasonal products. To ensure a steady supply, menu planners should frequently communicate with suppliers about the availability of fresh, seasonal produce. There is an ongoing debate in THE industry about the best months to eat mussels. Some argue for the 'r-months' (months with an 'r' in their name), while others disagree. From a sustainability perspective, late spring is the spawning season, so offering mussels on a menu during the 'r-months' is recommended.

Using a regional seasonal calendar can help determine when products are in season and assist in developing seasonal meal plans. Creating seasonal menus not only enhances variety but also effectively communicates these offerings to guests. Replacing frozen food with fresh, seasonal alternatives is less energy intensive. Organizing seasonal campaign weeks or festivals can raise guest awareness about the benefits of seasonal produce.

Organic and fair trade

Organic food is produced without synthetic chemicals or GMOs, using methods that promote ecological balance and biodiversity (Vigar et al, 2019). Organic agriculture is distinguished by its minimal use of external energy, reliance on natural self-regulation mechanisms, soil enrichment practices, closed resource cycles, natural plant protectants and a strong emphasis on animal welfare. These practices often result in healthier products compared to conventional farming, though they are more labour-intensive.

In many regions, inadequate regulation of social conditions, such as wages and working conditions, adversely affects small agricultural producers who lack the leverage to secure fair prices. This situation underscores the importance of fair trade, which aims to provide better trading conditions for marginalized producers and workers (Wheeler, 2012).

Incorporating fair and organic produce into your menu and effectively communicating this to customers can offer a competitive edge. Research by Wheeler (2012) indicates that organizations with fair trade certification tend to exhibit higher sustainable performance due to adherence to established standards. Certification schemes that evaluate criteria such as the percentage of locally sourced products, certified organic products, certified fair trade products and the avoidance of endangered species can significantly enhance an organization's sustainability profile (Seyfang, 2007).

Conventional farming practices often expose farm workers to harmful pesticides, herbicides and other toxins. In contrast, organic farming provides a healthier working environment, benefiting from cleaner air, water and soil (Lang and Lemmerer, 2019).

To develop a sustainable menu, priorities should include sourcing organic and fair trade products and buying directly from producers. These approaches should also be highlighted to customers. Studies show that consumers appreciate organic food and are willing to pay a premium for it. (For more information about certification, please refer to Chapter 5).

Menu layout

Designing a menu that promotes sustainable food choices is essential for reducing environmental impact. By strategically increasing the availability of vegetarian options, using effective language and highlighting sustainable practices, menu planners can influence customer choices while supporting their sustainable objectives. Parkin and Attwood (2022) offer several practical tips for designing a menu that promotes sustainable food choices:

1 Increase the availability of vegetarian dishes: Their research shows that a menu with 75 per cent presence of vegetarian dishes is the most effective in encouraging meat-eaters to choose a vegetarian option.

2 Placement of vegetarian symbols: Using a 'V' to denote vegetarian dishes to denote allergies rather than diet preference seems to be a more appropriate labelling, as it does not deter meat-eaters.

3 Shifting dietary choices: Prioritizing vegetarian dishes on a menu is essential for reducing the environmental impact of livestock production and achieving global greenhouse gas (GHG) emission reduction targets.

4 Predominantly vegetarian menus: Menus should predominantly feature vegetarian options while still offering a more limited range of meat options to satisfy customer preferences.

Menu language can also be an effective tool for encouraging sustainable food choices. Attwood et al (2019) found that altering the way in which dishes are described can lead to a 25 per cent increase in customers selecting vegetarian dishes. For instance, a message at the top of a menu emphasizing the environmental benefits of choosing vegetarian options, or highlighting sustainable practices such as local sourcing and reducing food waste, can encourage customers to choose plant-based options.

Another effective strategy is to make meat-free meals the default option, with a choice to add meat for an extra charge. For example, offering a black bean chilli as the main dish and allowing customers to add meat promotes not only sustainability but also inspires creativity in plant-based menu design.

Furthermore, Attwood et al found that terms like 'meat-free', 'vegan' or 'low-fat' can imply restriction or sacrifice. Using sensory adjectives like 'rich', 'crunchy' or 'creamy', to highlight flavour and texture seemed to have a less negative connotation. For example, renaming 'meat-free sausage and mash' to 'Cumberland-spiced veggie sausage and mash' increased sales by 76 per cent (Attwood et al, 2019).

The same study also showed that integrating plant-based dishes in the main menu rather than in a separate section doubled customer selection of vegetarian options from 6 per cent to 13 per cent. This strategy not only promotes sustainability but also enhances customer perception and choice.

To further enhance menu sustainability, another point of interest that could be included on the menu is the food miles of dishes.

Food miles (or kilometres)

Menus can also communicate food quality and operational values. According to Garnett (2010), the term 'food miles' associated with menu planning was introduced in the 1990s. Assessing sustainability through food miles, which measures the distance food travels from production to consumption along the supply chain (Van Passel, 2013), is one effective way to assess and communicate sustainability. (For more information on the supply chain please refer to Chapter 7). Table 8.2 helps explain sourcing impacts and encourages designing menus that minimize food miles, while promoting sustainability. To calculate how far food has travelled, a good starting point is www.foodmiles.com.

While food miles are a common measure, other factors like agriculture, processing, storage and shopping also contribute to the carbon footprint of food. Feeding the world is costly, consuming half of the Earth's habitable land. Food systems, encompassing production, processing, transport and consumption, account for one-third of human-caused emissions (Li et al, 2022; Yu et al, 2025). It is estimated that the carbon footprint of the global food transport system, emissions from transporting fertilizers, machinery, animal feed and food itself cause 3 billion tonnes of CO_2 emissions annually, 3.5-7.5 times higher than previous estimates. High-income nations, representing 12.5 per cent of the population, account for 52 per cent of international food miles and 46 per cent of emissions. While buying local food reduces emissions, the study suggests pairing it with eating seasonal produce and reducing meat consumption for greater impact.

The accuracy of food miles is often a contentious issue, with various perspectives and ongoing debates highlighting its complexity and the lack of consensus. Nevertheless, switching to more sustainable food choices can reduce emissions considerably, particularly in richer countries, as high emissions per head are in part due

Table 8.2 Calculating food miles for a sustainable menu

Objective	To help understand the concept of food miles and their impact on sustainability in menu planning.
Materials needed	Sample menu with a list of ingredients
	Access to the internet or a database for sourcing information
	Calculator
Instructions	
1. Introduction	Read the section on food miles in this chapter.
2. Menu selection	Choose your favourite restaurant. Go online and download their menu. Make sure the menu includes a variety of dishes with different ingredients.
	Pick one dish from each of the courses - one starter, one main, one dessert. A beverage can be included, such as coffee or bottled water.
3. Ingredient sourcing	Using the menu choices you have made, find out the recipe for each course and list all the ingredients.
	Research the origin of each ingredient and find out where the ingredients are commonly sourced (e.g. tomatoes from Italy, beef from Scotland).
4. Calculating food miles	Calculate the distance each ingredient travels from its source to the place where the menu is served (for example Clacton-on-Sea, Berlin, Mumbai), using online tools or maps to determine the distance in miles/km.
	Record the ingredients, their source and the calculated food miles
5. Analysis	Once all food miles are calculated, analyze the data to determine which dishes have the highest and lowest food miles.
	Think about what you have learned. What has surprised you? You might want to explore alternatives for sourcing ingredients locally.

to the wider food choices available in rich regions, such as North America, Australia, Europe and parts of Asia. This highlights the importance of rethinking food sourcing and menu planning to minimize environmental impact.

REAL-WORLD EXAMPLE Silo: A zero-waste blueprint

Rowen Halstead, Michelin-star chef and sustainable food advocate, uses the pioneering Silo restaurant to illustrate the practical application of zero-waste principles and the imaginative approach needed to redefine sustainability in food service.

'Waste is a failure of the imagination,' states Douglas McMaster, founder of Silo in London, the UK's first zero-waste restaurant. This philosophy, inspired by the idea that nature itself produces no waste, guides Silo's entire operation. McMaster echoes

Desmond Tutu's sentiment about needing to go upstream to solve problems – for Silo, this means fundamentally rethinking the food system, starting with how ingredients are sourced and utilized.

Achieving zero waste requires meticulous attention to detail and often means making things from scratch to avoid packaging. McMaster illustrates this with the example of a simple shortbread biscuit: flour must be milled from grain, and butter churned from cream delivered in reusable metal urns directly from the farm. This approach considers both flavour and environmental impact, making a zero-waste system logical. Silo's commitment has earned accolades, including a Michelin green star, proving that sustainability can be moulded around the restaurant's format, rather than being an afterthought.

This model challenges the conventional approach often seen in the struggling restaurant industry. It reconnects the kitchen with the food system, moving away from reliance on supermarkets that offer out-of-season produce and encourage bulk buying that leads to waste. Establishing direct connections with farmers is key; it fosters commitment, allows for menu planning based on seasonality, keeps creativity flowing, reduces waste and mirrors natural cycles.

McMaster argues that much restaurant food waste – estimated at 30 per cent of food purchased – stems from poorly designed menus striving for superficial 'perfection'. Cutting vegetables into perfect shapes often means discarding perfectly edible trim, peels, and edges. A waste-conscious menu, however, embraces creativity by finding uses for these 'imperfect' parts – incorporating trim into purees or stocks, dehydrating elements, or using fermentation. Silo, for example, uses fermentation extensively, transforming items like waste bread, cheese rinds and fish scraps into flavour-enhancing misos and garums, reducing food waste to just 0.5 per cent (with the remainder composted).

Silo's zero-waste philosophy extends beyond food. Wine bottles are recycled into light fittings and tableware, corks become flooring or staff shoe soles, and unavoidable plastic packaging is repurposed into items like chopping boards. This holistic approach provides a powerful blueprint.

Implementing such changes requires applying the 5 Rs: refuse, reduce, reuse, repurpose and recycle:

- **Refuse and reduce**: Critically assess ingredients. Can high-impact items like red meat be refused or reduced in favour of more sustainable local options like game? Explaining these choices often earns customer respect.
- **Reuse and repurpose**: This is vital for cutting waste and costs. Challenge the need for single-use garnishes; instead, repurpose trim or scraps creatively. If plate waste is high (an average 39 per cent of food served isn't eaten), consider smaller portions or fewer courses.

- **Recycle**: Experiment with recycling scraps into new dishes. This becomes a creative challenge for chefs.

The financial argument is compelling: food waste costs the UK hospitality industry an estimated £3.2 billion annually, yet every £1 invested in reducing waste saves around £7. As Anne-Marie Bonneau, the Zero-Waste Chef, notes, 'We don't need a handful of people doing zero waste perfectly. We need millions of people doing it imperfectly.' Small, consistent steps, inspired by pioneers like Silo, can create significant change, transforming industry standards dish by dish.

ACTIVITY 1: ZERO-WASTE MENU IMAGINATION

Objective

To apply zero-waste principles creatively to common kitchen byproducts, inspired by Silo's philosophy.

Instructions

1. **Choose two items** from the list below, which are often considered 'waste' in conventional kitchens:
   - vegetable peelings and trim (e.g. carrot peels, broccoli stalks, onion skins)
   - stale bread or bread crusts
   - used coffee grounds
   - cheese rinds (hard cheeses)
   - meat/fish bones or trim
2. For each of your chosen items, brainstorm and **briefly describe two or three innovative culinary uses** that transform this 'waste' into a valuable menu component (e.g. a garnish, a flavour enhancer, a stock, a new dish element).
3. Finally, **explain briefly how your ideas reflect Silo's philosophy** that 'waste is a failure of the imagination' and contribute to a zero-waste approach.

While resourcefulness in using every part of an ingredient is key, as highlighted by the Silo real-world example and the preceding activity, achieving truly sustainable food systems also involves looking at the bigger picture of dietary composition, as frameworks like the Planetary Health Diet illustrate.

The Planetary Health Diet

The Planetary Health Diet is a dietary approach designed to benefit both human and planetary health (Loken, Willett and Rockström, 2021; WHO, 2023). This dietary approach is largely derived from a multi-criteria approach to sustainable diets, drawing from the UN Food and Agriculture Organization's definition of food sustainability (Walker, DeMatteis and Lienert, 2021; Willett et al, 2019). See Figure 8.1.

Each criterion represents an aspect of sustainability that can be linked to a food product. While it is practically impossible for a single food item to meet all criteria, even satisfying one criterion can enhance sustainability. The more criteria a product meets, the closer it aligns with the ideals of environmental, social, and economic sustainability (Steffen et al, 2015).

This diet, the culmination of three years of research by the EAT-Lancet Commission, underscores the vital connection between humans' food choices and the health of the environment. It advocates for a balanced intake that shifts the world to healthier, tastier and more sustainable diets, realigns food system priorities for people and the planet, produces more of the right food from less, safeguards land and oceans and radically reduces food losses and waste. Considering the prediction that the global population will reach 10 billion by 2050, the Planetary Health Diet is based on the principle that food is a crucial link between human health and environmental health.

Beyond creatively minimising waste from the food we prepare, sustainable menu planning must also address the fundamental choice of ingredients and overall dietary balance. The Planetary Health Diet provides a framework connecting food choices to both human and planetary health.

Figure 8.1 The Planetary Health Diet

Fresh produce	50%
Whole grains	17%
Root vegetables	1%
Dairy products	4%
Animal-based protein	4%
Plant-based protein	12%
Unsaturated plant oils	9%
Refined sugars	3%

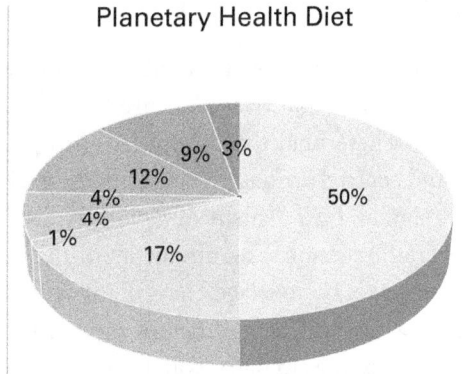

Adapted from EAT-Lancet Commission, 2021

The three-step sustainable menu: Planning, Designing, Presenting (PDP)

> **INDUSTRY VOICE: ROWEN HALSTEAD**
>
> Rowen Halstead, chef and sustainable food advocate, says: 'So much avoidable food waste is generated in kitchens simply through lack of knowledge or creativity. A sustainable menu isn't just about ingredients; it's a mindset. It starts with conscious planning: connect directly with local suppliers, embrace seasonality, and, crucially, design dishes that use the whole ingredient – root-to-fruit, nose-to-tail. Seeing "waste" like trim or peelings as opportunities sparks creativity and dramatically cuts costs. Understanding and implementing zero-waste practices isn't just ethical; it's vital for the future resilience and success of food businesses.'

A successful sustainable food and beverage policy requires careful menu planning and composition. It is crucial to meet client needs while meeting the bottom line and achieving the organization's sustainability goals. This complex task involves gradually introducing changes to the menu, considering all aspects of sustainability. While eliminating unsustainable elements like beef may not be feasible, there are ways to gradually reduce and replace such components with sustainable alternatives (Coskun, Hüseyin Genç and Coskun, 2023). Sustainable menu planning also involves considering water and energy usage during food preparation and minimising food waste. The most critical decision in a professional kitchen is the preparation and composition of menus.

Planning

The first step towards sustainable menus begins with the establishment of guiding principles and objectives that are customized to the specific facility, location and services. These objectives should aim to provide a food service that is sustainable, safe and centred around the customer's needs and preferences. They should facilitate the design and assessment of a standardized menu that caters to the dietary preferences and nutritional requirements of the customers. Similarly, procurement guidelines need to be in place that align with regional policies and national nutrition guides, while also being flexible enough to cater to a diverse clientele. The importance of multidisciplinary collaboration cannot be overstated when it comes to enhancing customer service (Höijer et al, 2020).

When planning the menu, the preferences of the clientele and the factors influencing food consumption should be considered. This involves striking a balance between controlling costs and maintaining quality and variety. Sustainability should be a primary concern right from the initial stages of menu planning. Before introducing a new menu, it is crucial to consult with customers, chefs, food service managers, nutritionists, kitchen staff and the sustainability officer (if such a role exists). Regular reviews of the established standards are necessary to ensure the continuous improvement of the food service (Sterling et al, 1996). This approach ensures a comprehensive and sustainable approach to menu planning and implementation.

Designing

The second step begins with defining the menu format, taking into account the constraints of cost and available resources. The production and distribution of the chosen menu will depend on the budget, kitchen facilities, storage spaces and the skills of the available staff. An à la carte menu, possibly combined with a cyclical (seasonal) menu of main dishes, could be the best way to meet the needs of different clientele types, ensure their satisfaction, and reduce food waste. The composition of the menus should be varied, with an emphasis on seasonal highlights and themed food weeks or months. Menu planners should consider using local, seasonal, organic, and fair trade products, and plan their purchases to avoid waste (Kallel, Kanoun and Dhouib, 2022).

When choosing ingredients, menu planners should prioritize produce and aim for a diverse range of fruits and vegetables across all courses. This allows for menu variation and adaptation to the season and availability of local food. Efficient resource use is also crucial. There should also be consideration regarding opting for energy-efficient cooking routines, such as à-la-minute preparation, and consider purchasing energy from renewable sources. Finally, the menu should be reviewed and modified regularly, at least twice a year or as needed in response to feedback. Similarly, changes should be made following fresh seasonal produce and to comply with established nutritional and sustainability criteria (Parkin and Attwood, 2022).

Presenting

The third step involves considering dish size and scope, reducing portion sizes at buffets, and emphasizing quality over quantity. It is about creating a culinary adventure with colourful, fresh ingredients. Highlight important dishes, arrange buffets to showcase sustainable foods, and offer attractive vegetarian and vegan alternatives. Menu planners should create special areas for organic food and local specialities. Play with colours and textures to make dishes stand out. Creating a pleasant dining atmosphere is crucial. There should also be considerations in regarding to decorating tables properly, present raw ingredients next to dishes to show food origins and train staff to recommend sustainable dishes.

It should also be ensured that guests understand the changes made and what goals are meant to be achieved. Inform them about the origin and seasonality of the food, the way it was produced and its nutritional values. Animal welfare and the protection of certain species should be emphasized. Using communication to create a positive image of the business and to make the menu unique is paramount (Paul and Paul, 2013).

A sustainable menu involves engaging all stakeholders. The clientele's satisfaction is key to a menu's longevity, and their feedback is crucial in both the creation and evaluation stages. Before introducing a new menu, it is essential to consult with all stakeholders, including customers, employees, suppliers, nutritionists and others. This helps consolidate the vision and facilitates necessary changes. When introducing sustainable principles, it is important to explain the sustainable direction the menu will take, encouraging positive support for habit changes (Huang, Hall and Chen, 2023). The post-implementation evaluation is vital for sustainable production and distribution, and departments involved should adopt an evaluation procedure to assess a menu's acceptability. This includes gathering and compiling information from satisfaction questionnaires, audits and feedback from staff. Finally, regular revisions of the menu, based on gathered data, are necessary for continuous improvement.

Building on these principles, the next step involves integrating sustainable operations within the food service department.

Bringing the sustainable menu to life: Operations

Developing a sustainable menu concept, as discussed in the previous sections, is only the first step. To bring this vision to life effectively and consistently requires aligning the day-to-day operations of the food and beverage department. The most well-designed sustainable menu will fail if operational practices – from personnel skills and training to equipment use and preparation processes – do not actively support its goals (Legrand, Sloan and Chen, 2017). This section explores how integrating sustainability into everyday operations is crucial for the successful implementation of a sustainable menu strategy.

Personnel requirements

In the context of sustainable menus, the personnel structure of the food and beverage department needs to be re-evaluated based on the requirements needed for the planning of sustainable menus (Aremu and Adepoju, 2022). This involves ensuring the necessary conditions for the whole food and beverage team to be able to deliver sustainable menus. It is therefore recommended that the THE organization carries out a skill assessment within the food and beverage department to consider the extent to which the team should undergo training to ensure effective food production and service delivery. In the case of food production (kitchen staff), this entails ensuring recipe preparation knowledge, dish preparation flexibility and skills in recognizing food freshness. In the case of food service delivery (service staff), this includes recipe knowledge, and service techniques.

Integrating sustainable practices may require operational changes and the provision of essential training for kitchen staff, service personnel and management. Achieving customer satisfaction is a shared responsibility that hinges on efficient stakeholder interactions. Therefore, tailored training in sustainable practices is crucial to encourage habit changes that support the planning of a sustainable menu. (For more details on training, please refer to Chapter 3).

Working environment: Equipment

In the context of sustainable menus, the selection of equipment plays a significant role. Equipment should be powered by sustainable energy sources, such as those derived from renewable sources, hydroelectricity, solar or wind power and come with certified reduced energy or water consumption (for more information about consumption reduction certification, please refer to Chapter 5). An industrial dishwasher and pot washer can save water and energy when the quantity of dishes is sufficient. It is important that the long-term reliability of equipment is established by an annual preventive maintenance schedule. Daily equipment maintenance, including cleanliness, clear air intake and good-condition door seals, is crucial to prevent energy wastage (Zaid, Jaaron and Bon, 2018).

While these items may be more expensive initially, they offer a good return on investment through energy savings (Sanderford, McCoy and Keefe, 2018). Combination ovens, which use convection and steam, often reduce cooking times and save energy. If large quantities of hot food need to be stored, consider using a rapid cooling unit. Therefore, working with the procurement department to establish criteria for sustainable purchases of both large and small equipment should be integrated into the department's sustainable procurement policy.

Waste management options should be discussed with the relevant departments to ensure suitable onsite waste management options, such as composting and recycling, are available. (For more information about food waste reduction techniques, please refer to Chapter 7).

One aspect of equipment that is often overlooked is the containers used when food is to be taken out of the property. It is not just take-away businesses that should be considering this, but all establishments. They should consider avoiding wax-lined cardboard packaging, as these are often not recyclable. For aluminium and plastic containers, their recyclability can depend on the recycling provider. Most plastic containers have a triangle-shaped logo with a number in the middle indicating the type of plastic and its level of recyclability. The food and beverage operation should check with the local municipality/authority to see which are recyclable.

Practical steps can further enhance sustainability and operational efficiency, such as prioritizing refrigeration over freezing which not only improves energy efficiency but also ensures the inclusion of fresh food on menus. Cold storage areas should be appropriately sized and doors properly closed to maintain optimal conditions. Additionally, optimizing freezer efficiency by sizing it to actual needs helps in reducing energy consumption. Implementing a barcode system for maintaining inventory for immediate needs and emergency reserves can streamline operations (Becerra, Mula and Sanchis, 2022). To reduce or eliminate paper waste, businesses should consider automating inventory management and ordering processes for electronic dispatch to suppliers.

These practices not only support sustainability but also set the stage for high operational processes in food production within sustainable food and beverage operations.

Operational processes

In the context of sustainable menus, operational processes in food preparation are key to maintaining close alignment with customers' demands. This involves producing food based on actual needs, adjusting quantities based on daily sales and leftovers, monitoring inventory to prevent waste, and restocking food as per requirements. This means that the quantities of food to be prepared should be adjusted based on daily sales and the amount of leftovers from previous meals.

Inventory management plays a crucial role in this process. By keeping a close eye on the inventory, one can avoid wastage and ensure that the food items are always available as per the requirements. This involves restocking food items promptly, based on the actual need (Salama et al, 2022). This approach not only contributes to sustainability but also enhances the overall customer experience by ensuring the availability of fresh and appealing food options. Effective inventory management techniques are vital for this sustainability link; examples include strictly applying 'First-In, First-Out' (FIFO) stock rotation to minimize spoilage, utilizing inventory software or regular manual checks to track usage patterns accurately, identifying slow-moving items promptly to adjust orders or feature them creatively, and ensuring clear communication between kitchen and service teams about stock levels.

Developing a sustainable menu requires a holistic approach that integrates optimized food production and delivery, efficient personnel management, sustainable equipment selection and well-defined operational processes. Implementing these practices not only supports environmental sustainability but also improves customer satisfaction.

Networks

While the importance of local food is widely recognized, sourcing it can be challenging due to inadequate logistics, infrastructure and communication. Small-scale suppliers and local producers may also be unable to meet demand. Therefore, establishing robust local networks is essential for a sustainable food policy. Stein and Santini's (2022) three-stage process to networking is a useful starting point:

Stage 1: Define values and objectives

The business should establish a vision and mission that reflect what it stands for and aims to achieve in relation to sustainable practices. This facilitates the research the business should undertake to identify local producers and their products. It is important to identify who stakeholders are. While a business outlook may tend to view other business of a similar type as competition, there is great advantage in sharing knowledge. Potential network members could include representatives from hotels, restaurants, bars, as well as local producers and suppliers such as organic farmers. Networking with other food service managers, both regionally and nationally, can facilitate the search for sustainable products and local suppliers. The network may also include public institutions such as food cooperatives, non-governmental associations and local councils.

Stage 2: Clarify offerings

The business needs to have a clear idea of what it is offering. This will direct how to build the network around specific products. In this respect, identifying local producers and their offerings is crucial, as customers are often willing to pay a premium for authentic, local products. By gathering detailed information about these producers and their specialities, marketing efforts can be enhanced by making regional specialties more appealing and unique. Establishing connections with local producers is the first step towards creating reliable supply networks and diversifying suppliers can reduce dependence on single retailers.

Stage 3: Engage stakeholders

Ensure all key stakeholders within the business understand the benefits of local networks. These advantages can be made visible through regular agenda items at meetings and focused professional development. Including external stakeholders in meetings and training will provide employees with a better understanding of the complexities of sourcing. Involving all stakeholders in the food chain is also part of this step as awareness raising about sustainable food criteria can be achieved (Coskun, Hüseyin Genç and Coskun, 2023). It is counterproductive to build strong external networks if colleagues and employees are not convinced of the direction. All stakeholders need to be clear about their role as suppliers and buyers within the network. Ensuring a system of coordination and transparency through regular meetings can help build mutually beneficial, trusting relationships.

By following Stein and Santini's three-stage process, the formation of a network with other local businesses will enable a business to support its sustainable aspirations by underpinning the business's values and objectives, clarifying every stakeholder's role and business offer, and ensuring ongoing knowledge sharing and appropriate contact building.

The sustainable menu checklist in Table 8.3 provides practical steps to integrate these principles into operations, ensuring a comprehensive approach to sustainability.

Table 8.3 Sustainable menu checklist

A menu when **planning** should be:	A menu when **presenting** should be:
• Seasonal: The menu changes seasonally.	• Balanced: The menu on display is refilled more frequently rather than looking excessively opulent.
• Special: The menu hosts special themed local food, such as an Onion Festival (in an onion producing areas).	• Enticing: The menu makes sustainable choices appealing.
• Local: The menu highlights local production.	• Varied: Menu buffets prioritize vegetarian and vegan alternatives.
• Fresh: The menu prioritizes freshly prepared dishes, reducing any processed foods.	• Informative: The menu explains and informs its guests about the sustainable credentials of food.
• Social: The menu considers organic and fair trade options.	• Clear: The menu does not overwhelm customers with information.
• Savvy: The menu adopts energy-efficient cooking methods.	• Sincere: The menu projects a trustworthy impression.
• Green: The menu prioritizes a variety of fruits and vegetables in their recipes.	

Conclusion

This chapter has established the pivotal role menus play in advancing sustainability within the THE sector. From understanding the strategic importance of ethical sourcing, seasonality, and minimizing food miles to applying frameworks like the Planetary Health Diet and the three-step Planning, Designing Presenting approach, it is clear that menu development is a powerful lever for change. Integrating these principles into daily operations and building strong local networks further supports the creation of food offerings that are not only appealing to customers but also contribute positively to environmental health, and economic resilience.

Understanding the basics of sustainable menu development and the importance of ethical sourcing is crucial for anyone thinking about developing a sustainable menu. The following practical activity is designed to reinforce the key concepts discussed in the chapter. Identifying the correct answers will deepen the comprehension of sustainable practices and ethical considerations, ensuring well-preparedness to implement these principles. This hands-on approach not only solidifies theoretical knowledge but also encourages critical thinking and attention to detail, essential skills for any sustainability-focused professional.

ACTIVITY 2: EVALUATING SUSTAINABLE MENU ACTIONS

Below is a list of actions when thinking about developing a sustainable menu. Some are true, some are false. Identify which points are true based on the principles of sustainable menu development.

	Action	T	F		Action	T	F
1	Serve fresh, appetizing food.			10	Offer recipes of varied cultural origins representing the population you serve.		
2	Use recipes approved by a tasting panel representing the eaters in the facility.			11	Undertake to learn the cultural importance of foods eaten by the people in the local community in your community.		
3	Serve regional fruits and vegetables when they are in season.			12	Reduce the amount of red meats used (and increase the use of proteins with a lesser environmental impact).		

(continued)

(Continued)

	Action	T	F		Action	T	F
4	Offer fair trade products when possible.			13	Cut waste by using surplus and adjusting serving sizes.		
5	Serve food containing as many additives and preservatives as possible.			14	Offer bottled water as the sole basic beverage.		
6	Reduce the use of processed foods.			15	Limit highly processed foods.		
7	Serve shellfish and fish from sustainable fishing or aquaculture.			16	Choose protein foods that come from red meat more often.		
8	Compost and recycle waste.			17	Serve regional and seasonal produce.		
9	Serve animal products from animals raised with routine antibiotics.			18	Involve employees in the process of making the menu more sustainable.		

KEY TAKEAWAYS

- Understanding sustainable menus:
    - A sustainable menu involves promoting environmentally friendly and sustainable food choices.
    - It is crucial to recognize the importance of sustainable menu design as it not only supports environmental stewardship, but also aligns with economic viability and social responsibility.
    - It is key to contributing to long-term profitability and customer retention.
- Food miles:
    - Food miles measures the distance food travels from production to consumption.
    - Awareness of food miles helps in understanding the environmental impact of transporting food, and encourages sourcing ingredients locally to reduce carbon emissions.

- Three-step approach:
  - The chapter provides the instrumental knowledge to integrate sustainability: menu- planning, designing and presenting.
  - The chapter underscores the importance of aligning menu items with sustainability goals and customer preferences.
- Real-world examples and regulatory framework:
  - By studying real-world scenarios and the regulatory framework, you can identify effective strategies and common pitfalls in implementing sustainable menus.
  - This knowledge helps in understanding the practical application of sustainability principles supporting ethical sourcing and environmental responsibility.
- Tools and strategies:
  - Tools and strategies are included to help:
    - guide menu planners in incorporating sustainable practices
    - evaluate the sustainability of menu items and sourcing practices
    - provide a structured approach to planning, designing and presenting sustainable menus
    - promote sustainable practices in food and beverage services
    - foster collaboration with local suppliers and stakeholders
    - continuously assess and refine sustainability initiatives

REFLECTIVE QUESTIONS

1. What is a sustainable menu and why is it important in the food and beverage industry?
2. How can a menu influence consumer choice and promote sustainability?
3. Reflecting on the three-step approach (Planning, Designing, Presenting) for sustainable menus, what do you consider the biggest practical challenge or key opportunity for a THE business when implementing these stages?
4. What are the principles of the Planetary Health Diet, and why is it relevant to sustainable menu planning?

Further resources

SRA (2023) How to encourage sustainable choices through the language on your menu, The Sustainable Restaurant Association, 26 March, https://thesra.org/news-insights/insights/how-to-encourage-sustainable-choices-through-the-language-on-your-menu/ (archived at https://perma.cc/M3CF-LL3B)

UCONN (.n.d.) The principles of healthy, sustainable menus, Dining Services, University of Connecticut, https://dining.uconn.edu/the-principles-of-healthy-sustainable-menus/ (archived at https://perma.cc/Z8MH-87FD)

WWF (.n.d.) Catering for sustainability. Making the case for sustainable diets in foodservice, https://assets.wwf.org.uk/downloads/wwf_catering_full_report.pdf (archived at https://perma.cc/VZX9-ZTHL)

References

Aremu, A B and Adepoju, O O (2022) Green human resource management practices and employee retention: An empirical evidence from Nigerian foods and beverages industry, *International Journal of Innovative Social Sciences & Humanities Research*, **10** (4), pp 103–12

Attwood, S, Chesworth, S J and Parkin, B L (2020) Menu engineering to encourage sustainable food choices when dining out: An online trial of priced-based decoys, *Appetite*, **149**, p 104601, https://doi.org/10.1016/j.appet.2020.104601 (archived at https://perma.cc/S9SQ-FJHE)

Attwood, S, Voorheis, P, Mercer, C, Davies, K and Vennard, D (2019) Playbook for guiding diners toward plant-rich dishes in food service, World Resources Institute, 7 January, www.wri.org/research/playbook-guiding-diners-toward-plant-rich-dishes-food-service (archived at https://perma.cc/T6ZW-PM4M)

Becerra, P, Mula, J and Sanchis, R (2022) Sustainable inventory management in supply chains: Trends and further research, *Sustainability*, **14** (5), p 2613

Brune, S, Knollenberg, W, Barbieri, C and Stevenson K (2023) Towards a Unified Definition of Local Food, *Journal of Rural Studies*, **103**, p 103135, https://doi.org/10.1016/j.jrurstud.2023.103135 (archived at https://perma.cc/6D5B-UBBT)

Coskun, A, Hüseyin Genç, H U and Coskun, A (2023) How sustainable is your menu? Designing and assessing an interactive artefact to support chefs' sustainable recipe-planning practices. In Proceedings of the 6th ACM SIGCAS/SIGCHI Conference on Computing and Sustainable Societies, pp 90–98

Feldmann, C and Hamm, U (2015) Consumers' perceptions and preferences for local food: A review, *Food Quality and Preference*, **40**, pp 152–64

Filimonau, V and Krivcova, M (2017) Restaurant menu design and more responsible consumer food choice: An exploratory study of managerial perceptions, *Journal of Cleaner Production*, **143**, pp 516–27

Garnett, T (2010) The food miles debate: Is shorter better? In A McKinnon (ed.) *Green Logistics: Improving the environmental sustainability of logistics*, Kogan Page, London

Hoarau-Heemstra, H, Wigger, K, Olsen, J and James, L (2023) Cruise tourism destinations: Practices, consequences and the road to sustainability, *Journal of Destination Marketing & Management*, **30**, p 100820

Höijer, K, Lindö, C, Mustafa, A, Nyberg, M, Olsson, V, Rothenberg, E, Sepp, H and Wendin, K (2020) Health and sustainability in public meals – an explorative review, *International Journal of Environmental Research and Public Health*, **17** (2), p 621

Huang, Y, Hall, C M and Chen, N (2023) The sustainability characteristics of Michelin green star restaurants, *Journal of Foodservice Business Research*, **28** (2), pp 219–44, https://doi.org/10.1080/15378020.2023.2235258 (archived at https://perma.cc/G6A3-53EL)

Imperfect Foods (.n.d.) What makes produce imperfect? Food Waste Movement, https://blog.imperfectfoods.com/what-makes-produce-imperfect/ (archived at https://perma.cc/CW7N-EWFN)

Ip, M M H and Chark, R (2023) The effect of menu design on consumer behavior: a meta-analysis, *International Journal of Hospitality Management*, **108**, p 103353

Kallel, D, Kanoun, I and Dhouib, D (2021) The menu planning problem: A systematic literature review. In *International Conference on Intelligent Systems Design and Applications*, Springer International Publishing, Cham

Kim, M J and Hall, C M (2020) Can sustainable restaurant practices enhance customer loyalty? The roles of value theory and environmental concerns, *Journal of Hospitality and Tourism Management*, **43**, pp 127–38

Lang, M and Lemmerer, A (2019) How and why restaurant patrons value locally sourced foods and ingredients, *International Journal of Hospitality Management*, **77**, pp 76–88

Legrand, W, Sloan, P and Chen, J S (2017) *Sustainability in the Hospitality Industry: Principles of sustainable operations*, Routledge, Abingdon

Lera, D (2019) Food sales. In A Martin (ed.) *The Practical Guide in Understanding and Raising Hotel Profitability*, Routledge, Abingdon

Li, Q, Sheng, B, Huang, J, Li, C, Song, Z, Chao, L, Sun, W, Yang, Y, Jiao, B, Guo, Z and Liao, L (2022) Different climate response persistence causes warming trend unevenness at continental scales, *Nature Climate Change*, **12** (4), pp 343–49

Loken, B, Willett, W and Rockström, J (2021) Food, planet, health: healthy and sustainable diets for 10 billion people. In *World Scientific Encyclopaedia of Climate Change: Case studies of climate risk, action, and opportunity. V2*, pp 107–17

Parkin, B L and Attwood, S (2022) Menu design approaches to promote sustainable vegetarian food choices when dining out, *Journal of Environmental Psychology*, **79**, p 101721

Paul, G and Paul, S (2013) Proposal for a novel computerized menu-presentation interface for restaurants. In *Proceedings of the 11th Asia Pacific Conference on Computer-Human Interaction – APCHI '13*, pp 119–22

Rockström, J, Gupta, J, Qin, D, Lade, S J, Abrams, J F, Andersen, L S, Armstrong McKay, D I, Bai, X, Bala, G, Bunn, S E and Ciobanu, D (2023) Safe and just Earth system boundaries, *Nature*, **619** (7968), pp 102–11

Salama, W, Nor El Deen, M, Albakhit, A and Zaki, K (2022) Understanding the connection between sustainable human resource management and the hotel business outcomes: Evidence from the green-certified hotels of Egypt, *Sustainability*, **14** (9), p 5647

Sanderford, A R, McCoy, A P and Keefe, M J (2018) Adoption of energy star certifications: Theory and evidence compared, *Building Research & Information*, **46** (2), pp 207–19, https://doi.org/10.1080/09613218.2016.1252618 (archived at https://perma.cc/RM74-T2BM)

Schmitt, E, Galli, F, Menozzi, D, Maye, D, Touzard, J M, Marescotti, A, Six, J and Brunori, G (2017) Comparing the sustainability of local and global food products in Europe, *Journal of Cleaner Production*, **165**, pp 346–59

Seyfang, G (2007) Growing sustainable consumption communities: The case of local organic food networks, *International Journal of Sociology and Social Policy*, **27** (3/4), pp 120–34

Smith, A and MacKinnon, J B (2009) *The 100-Mile Diet: A year of local eating*, Vintage Canada, Toronto

Steffen, W, Richardson, K, Rockström, J, Cornell, S E, Fetzer, I, Bennett, E M, Biggs, R, Carpenter, S R, De Vries, W, De Wit, C A and Folke, C (2015) Planetary boundaries: Guiding human development on a changing planet, *Science*, **347** (6223), p 1259855

Stein, A J and Santini, F (2022) The sustainability of 'local' food: A review for policymakers, *Review of Agricultural, Food and Environmental Studies*, **103** (1), pp 77–89

Sterling, L, Petot, G, Marling, C, Kovacic, K and Ernst, G (1996) The role of common sense knowledge in menu planning, *Expert Systems with Applications*, **11** (3), pp 301–08

UKHospitality (2022) Environmental sustainability guide: For SME in the hospitality sector, https://app.sheepcrm.com/ukhospitality/member-only-documents/environmental-sustainability-guide/ (archived at https://perma.cc/GEX2-9TKY)

Van Passel, S (2013) Food miles to assess sustainability: A revision, *Sustainable Development*, **21** (1), pp 1–17

Vigar, V, Myers, S, Oliver, C, Arellano, J, Robinson, S and Leifert, C (2019) A systematic review of organic versus conventional food consumption: Is there a measurable benefit on human health? *Nutrients*, **12** (1), p 7

Walker, C, DeMatteis, L and Lienert, A (2021) Selecting value chains for sustainable food value chain development – Guidelines, FAO, https://openknowledge.fao.org/items/197d1311-b31e-4203-9aac-53101ac455de (archived at https://perma.cc/L7GS-QFZ4)

Wansink, B and Love, K (2014) Slim by design: Menu strategies for promoting high-margin, healthy foods, *International Journal of Hospitality Management*, **42**, pp 137–43

Wheeler, K (2012) *Fair Trade and the Citizen-Consumer: Shopping for justice?* Springer, New York

WHO (2023) The State of Food Security and Nutrition in the World 2023: Urbanization, agrifood systems transformation and healthy diets across the rural-urban continuum, FAO, https://openknowledge.fao.org/items/445c9d27-b396-4126-96c9-50b335364d01 (archived at https://perma.cc/NXJ2-CVLK)

Willett, W, Rockström, J, Loken, B, Springmann, M, Lang, T, Vermeulen, S, Garnett, T, Tilman, D, DeClerck, F, Wood, A and Jonell, M (2019) Food in the Anthropocene: The EAT-Lancet Commission on healthy diets from sustainable food systems, *The Lancet*, **393** (10170), pp 447–92

Yu, W, Liu, F, Jiao, X, Fan, P, Yang, H, Zhang, Y, Li, J, Chen, J and Li, X (2025) Human-induced NP imbalances will aggravate GHG emissions from lakes and reservoirs under persisting eutrophication, *Water Research*, **276**, p 123240

Zaid, A A, Jaaron, A A and Bon, A T (2018) The impact of green human resource management and green supply chain management practices on sustainable performance: An empirical study, *Journal of Cleaner Production*, **204**, pp 965–79

9 | Greenwashing

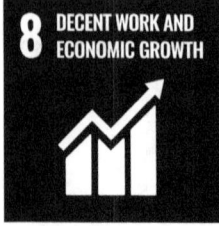   

Systems thinking competency | Critical thinking competency | Strategic competency | Normative competency

CHAPTER AIM

The aim of this chapter is to encourage a practical understanding of greenwashing, its implications within the tourism, hospitality and events (THE) industry, and the associated legislation designed to combat it. Emphasis is placed on aligning the understanding of greenwashing with key global frameworks, including the Sustainable Development Goals (SDGs), particularly SDG 12 (Responsible Consumption and Production), SDG 13 (Climate Action), SDG 16 (Peace, Justice and Strong Institutions), and SDG 8 (Decent Work and Economic Growth), as misleading claims undermine progress across these areas.

The chapter provides guidance on recognizing and avoiding instances of greenwashing, thereby supporting the development of key sustainability competencies such as critical thinking, normative competency, systems thinking and strategic competency. Through exploring definitions, regulations, real-world examples and avoidance strategies, learners are encouraged to critically evaluate communication practices and identify pathways towards more authentic and responsible business operations.

LEARNING OUTCOMES

Upon completion of this chapter, you will be able to:

- understand the concept of greenwashing and its implications in THE
- identify key legislation involving greenwashing.
- recognize instances of and understand how to avoid greenwashing.
- critically evaluate the ethical considerations surrounding greenwashing

KEY WORDS

Advertising, corporate communication, green claims, green hushing, greenwashing, trust.

Introduction

This chapter explores the critical issue of greenwashing within the tourism, hospitality and events (THE) industry, examining the practice of misleading consumers and stakeholders about environmental or sustainability performance. It begins by establishing core definitions of greenwashing and related terminology. The chapter then reviews key EU and UK regulations and the role of regulatory bodies in ensuring truthful environmental claims. Subsequently, it delves into common greenwashing tactics, utilizing the seven sins framework as a tool for identification, and analyzes real-world examples through a detailed case study and specific instances observed within THE operations. Finally, the chapter equips readers with practical tips and actionable strategies to effectively identify and avoid greenwashing practices in their own future roles, fostering more transparent and trustworthy communication essential for genuine sustainability. An activity is included to consolidate learning and encourage critical debate on the topic

According to Orange and Cohen (2010), the term 'greenwashing' was coined in 1986 when environmentalist Jay Westerveld observed hotels promoting towel reuse as an eco-friendly initiative, despite it being a cost-saving measure. The term refers to companies portraying themselves as environmentally friendly for profit. The concept gained momentum from campaigns like Keep America Beautiful in 1953 and the environmental movement of the mid-1960s. Despite the US Federal Trade Commission's Green Guidelines in 1998, deceptive environmental claims persist. A

2024 EU study found over 50 per cent of such claims to be misleading or unfounded. Efforts to expose greenwashing include publications by the Advertising Standards Authority (ASA), Competition and Markets Authority (CMA), Greenpeace and the Greenwashing Academy Awards.

A universally accepted definition of greenwashing is a contested terrain. This chapter will use the United Nations definition as '…the misleading [sometimes intentional, most times unintentional] approach all companies including THE have taken to demonstrate commitment to sustainability' (UN, .n.d. a).

Greenwashing is closely related to SDG 12, which focuses on ensuring sustainable consumption and production patterns. Greenwashing can undermine efforts to achieve this goal by misleading consumers and promoting unsustainable practices as environmentally friendly. In the THE industry greenwashing can take a multitude of forms and can permeate all stages of the supply chain. As the THE sector strives to meet the demands for sustainable practices, ethical communication becomes ever more paramount. Notably, greenwashing can often be an unintentional act.

As stated earlier, greenwashing describes a company or destination's deceptive marketing strategy, when they present themselves as environmentally friendly or sustainable, but their actions contradict these claims. This can involve unsubstantiated claims, misleading tactics, or focusing on minor practices while neglecting significant environmental impacts. According to the European Supervisory Authorities (ESAs), 'greenwashing is a practice where sustainability-related statements, actions, or communications do not clearly and fairly reflect the underlying sustainability profile of an entity, a financial product, or financial services. This practice may be misleading to consumers, investors, or other market participants.' (European Securities and Market Authority, 2024)

Consumers of tourism are becoming increasingly aware of THE businesses' environmental claims. Yet, greenwashing can attract tourists seeking eco-conscious options by creating a false image of sustainability. Tourists believe they are making responsible choices, while their actions might contribute to environmental degradation. As sustainability awareness grows, the tourism industry responds with sustainability initiatives. However, not all claims are genuine. Greenwashing tactics range from vague language and buzzwords to exaggerating or outright lying about sustainability efforts. This not only misleads consumers but also undermines the efforts of truly committed businesses and destinations.

The risks of greenwashing

Public concern for the environmental practices of businesses is growing. As tourists value sustainable operations, THE companies face increasing pressure to meet stakeholder demands (Choudhury, Islam and Sujauddin, 2024). Aware of the importance of image, and reputation, some THE organizations might be tempted to exaggerate, or inadvertently misinform their customers about their environmental efforts to create a positive public image. This environmentally friendly 'positive green communication' can be deceiving when a company's actions do not match its claims.

Driven by a growing concern for the environment, consumers are increasingly willing to pay a premium for sustainable products and services (De Freittas-Netto et al, 2020). This heightened awareness creates a vulnerability to greenwashing tactics. Companies may invest heavily in marketing themselves as 'green' without actually implementing sustainable practices in their production or operations. This deception can trick consumers into believing they are making environmentally conscious choices. By misleading the public to believe that a company or other entity is doing more to protect the environment than it is, greenwashing can promote false solutions to the climate crisis, distracting from and delaying concrete and credible action. Similarly, in some other cases, regulators have indeed demanded levels of evidence that are either unattainable or unhelpful. This overreach can lead to 'greenhushing' (see later in this chapter) or green inaction, where companies refrain from making any environmental claims to avoid scrutiny. Such outcomes are arguably worse, as they stifle transparency and hinder progress towards genuine sustainability.

Regulatory bodies

In response to an increase in greenwashing, EU and UK regulatory bodies expanded their remit to include greenwashing in their standards' frameworks. Bodies such as the International Consumer Protection and Enforcement Network (ICPEN), the European Commission (EC) and the Competition and Markets Authority (CMA) use consumer protection laws and advertising standards to keep companies honest when declaring their environmental impact. The THE industry, regardless of whether business operations are local or global, is obligated to abide by these regulations.

At a global level, the ICPEN provides guidance about greenwashing claims, representing consumer protection authorities from 70 countries. Established in 1992, the ICPEN aims to foster collaboration on issues of consumer protection, particularly concerning cross-border purchases of goods and services. The ICPEN is committed to safeguarding consumers globally by promoting and enabling proactive measures among its members. This involves the exchange of information about market trends and regulatory practices, as well as coordinating efforts to address market issues.

For instance, ICPEN collaborates with national or transnational authorities, or consumer groups such as Consumer International (CI), to align standards globally. For example, ICPEN addressed misleading claims by travel agencies about eco-friendly accommodations through a global review, finding that 40 per cent of green claims were potentially misleading (CMA, 2021a). This led to increased scrutiny and the development of guidelines to ensure transparency in environmental marketing in THE among other businesses. More recently the CI has contributed to the EU 2022 proposal, adopted in 2024, on empowering consumers for the green transition (European Union Agency for Fundamental Rights, 2024).

Regulations in the EU

The European Commission adopted a 'Proposal for a Directive on Green Claims' on 22 March 2023. This proposal aims to combat misleading environmental claims and ensure consumers receive reliable, comparable and verifiable environmental information. The proposed directive establishes clear criteria for companies to demonstrate the validity of their environmental claims. Additionally, claims and labels will require verification by independent, accredited bodies. Furthermore, new regulations aim to ensure environmental labelling schemes are robust, transparent and reliable. The Green Claims Directive, once implemented by individual member states, will introduce stricter controls for companies, including regulations on published materials and reported information. Member states can impose fines of up to 4 per cent of a company's profits and deny access to contracts for companies making unsubstantiated environmental claims (European Commission, .n.d.).

The 'Proposal for a Directive on Green Claims' exemplifies the growing focus on tackling greenwashing. The adopted proposal requires stricter evidence-based environmental claims from businesses in the EU region. THE businesses face fierce competition in showcasing their 'green' credentials. This is particularly pertinent when considering environmental activists and consumers can swiftly expose greenwashing through social media, damaging a business's reputation. This proposed legislation serves as a reminder that a strong sustainability strategy alone is insufficient. Businesses require data-driven approaches and well-informed communication teams to substantiate claims and avoid accusations of greenwashing.

The European Commission (.n.d.) reports on their website that:

- roughly half (53 per cent) of green claims provide misleading, vague or unsubstantiated information
- 40 per cent lack any supporting evidence
- half of all green labels offer weak or non-existent verification
- 230 sustainability labels and 100 green energy labels in the EU have vastly different levels of transparency

These findings highlight the urgent need for stricter regulations to combat greenwashing and enlighten consumers.

Once implemented, each EU member state will be required to transpose the directive into national law, potentially with even stricter regulations. This will significantly decrease the opportunity for companies to disseminate misleading information. Consumers will also gain new tools to identify and report instances of greenwashing to relevant authorities. The significance of this change cannot be overstated. It would be the only legal area requiring pre-approval of claims. While foods can be marketed without any checks, making a green claim would necessitate verification. This disparity highlights the level of scrutiny applied to environmental claims compared to other product categories.

Under the new EU law being implemented, it is illegal to make a generic environmental claim unless the trader can demonstrate recognized excellent environmental performance relevant to the claim. Examples of such generic environmental claims include terms like 'environmentally friendly', 'eco-friendly', 'green', 'nature's friend', 'ecological', 'climate friendly', 'gentle on the environment', 'carbon friendly', 'energy efficient', 'biodegradable', 'biobased' or similar statements that imply superior environmental performance. Recognized excellent environmental performance can be demonstrated by compliance with Regulation (EC) No 66/2010, officially recognized EN ISO 14024 ecolabelling schemes in the Member States, or by meeting top environmental performance standards for specific characteristics in accordance with other applicable Union laws, such as achieving class A under Regulation (EU) 2017/1369 of the European Parliament and of the Council.

Regulations in the UK

In the UK, advertising across all media is primarily regulated by the Competition Market Authority (CMA) and the Advertising Standards Authority (ASA). These entities ensure that advertising adheres to principles of legality, decency, honesty and truthfulness by enforcing codes set by the Committee of Advertising Practice (CAP) (see Figure 9.1).

The CMA functions as the UK's primary authority on competition and consumer protection, specifically through its Guidance on Environmental Claims on Goods and Services. This guidance explicitly states that while consumer protection law permits environmental claims, it strictly forbids misleading consumers (CMA, 2021). The CMA's guidance is dual-purpose: firstly, it empowers consumers with clear, truthful and accurate information to make informed choices aligned with their environmental values; secondly, it promotes fair competition, enabling businesses with authentic environmental practices to compete effectively. The CMA, tasked with safeguarding consumers from misleading practices, ensures businesses comply with consumer protection laws, including those specifically related to environmental claims. It collaborates closely with local Trading Standards Officers, who enforce consumer protection regulations at the local level. While the

Figure 9.1 List of UK regulators

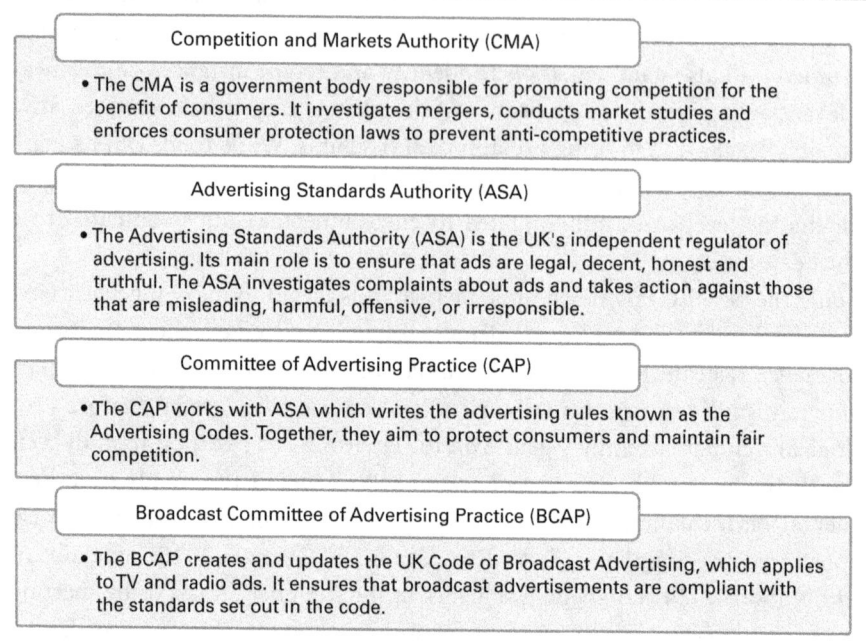

CMA addresses broad national and systemic issues, local enforcement is managed by Trading Standards Officers, ensuring consistent compliance nationwide.

Central to advertising regulation is the CAP, which formulates the UK Advertising Codes enforced by the ASA. The ASA's role is to investigate complaints and ensure compliance, thereby maintaining standards that prevent advertisements from being misleading, harmful or offensive. The ASA also manages evolving advertising practices, such as online behavioural advertising and influencer marketing, ensuring contemporary methods adhere to high regulatory standards.

CMA's detailed guidance is crucial for businesses in THE industries, helping them understand and adhere to legal obligations surrounding environmental claims. These guidelines apply to claims made about products, services, brands or specific aspects or components thereof. Businesses are urged to review the CMA guidance thoroughly to avoid inadvertently misleading consumers.

The core principles outlined in the CMA's guidance are as follows:

- claims must be truthful and accurate
- claims must be clear and unambiguous
- claims must not omit or hide important relevant information
- comparisons must be fair and meaningful
- claims must consider the full life cycle of the product or service
- claims must be substantiated

An example of language open to interpretation in the list is 'comparisons must be fair and meaningful'. The terms 'fair' and 'meaningful' can be subjective and interpreted differently depending on the context and perspective, especially in a court of law (Baude and Sachs, 2016).

To effectively comply with these standards, THE businesses must proactively audit their environmental claims, ensuring accuracy and clarity. This involves removing misleading or unsubstantiated claims, revising existing claims for veracity, and gathering comprehensive, verifiable data supporting any environmental assertions. Transparency with consumers through clear and comprehensive information is fundamental, enabling informed consumer choices and preventing allegations of misleading advertising. Misleading claims, particularly those selectively highlighting specific eco-friendly aspects without addressing broader environmental impacts, are subject to regulatory scrutiny. For instance, a business claiming 'eco-friendly' status due to energy-efficient lighting, but failing to disclose that it relies on non-renewable energy, could be flagged as misleading.

Failing to comply with these regulatory standards can lead to significant consequences. The CMA holds authority to initiate legal proceedings against businesses in breach of consumer protection laws. In some cases, companies may also be required to compensate affected consumers. Additionally, the ASA can act against advertisements violating the UK Code of Non-Broadcast Advertising and Direct and Promotional Marketing and UK Code of Broadcast Advertising (CAP and BCAP codes), mandating their modification or withdrawal.

The CMA shares responsibility for consumer protection enforcement with the ASA under the Consumer Protection from Unfair Trading Regulations 2008. Specifically, the CMA oversees adherence to the CAP and BCAP codes. When identifying advertising-related consumer protection concerns, the CMA may refer these issues directly to the ASA for resolution, maintaining consistency across regulatory actions. The CMA also coordinates enforcement efforts with other regulatory bodies and sectoral regulators. When investigating environmental claims, these regulators collaboratively determine the most suitable authority to handle each case. This collaborative approach ensures coordinated and comprehensive enforcement of regulations. Businesses may also face legal action initiated directly by consumers misled by inaccurate or false claims. These consumers have legal recourse through the courts to seek compensation for breaches of consumer protection law.

Table 9.1 outlines examples from the CMA guidance highlighting the potential pitfalls of misleading environmental claims.

These examples underscore the complexity and importance of clarity in environmental claims. Businesses must clearly disclose all relevant environmental impacts associated with their products or services, ensuring transparent and accurate consumer communication. Including disclaimers or small print addressing aspects not explicitly covered in a claim can mitigate potential misleading assertions. Additionally,

Table 9.1 Three examples of greenwashing pitfalls

1. Misleading eco-claim	2. Unclear recycling claim	3. Deceiving compostable claim
The company boasts a '33% lower carbon impact' product. This excludes transportation emissions, a major contributor to the product's footprint.	A product labelled 'recyclable' confuses consumers. It's unclear if the entire product or just the packaging can be recycled.	'Compostable' cup lacks details. It will not decompose in home bins, requiring industrial composting most consumers lack access to.
Consumers are misled by headline claims.	Since only the packaging is recyclable (a small part), this claim misleads consumers.	Poor labelling is misleading eco-conscious consumers.
Deceptive footprint: A product's advertised "lower carbon impact" focuses on reduced production emissions. However, this ignores transportation emissions, potentially the largest source of its environmental impact.	Clarifying recyclability: To avoid misleading consumers, a product labelled 'recyclable' should specify if it refers to the whole product or just the packaging. Clearly stating 'recyclable packaging' ensures accurate information for environmentally conscious customers.	Transparency in composting: A 'compostable' cup should clearly state its limitations. Specify 'industrial composting only' to avoid misleading consumers who lack access to such facilities. Transparency builds trust with environmentally conscious customers.

Adapted from CMA, 2021b

the upcoming Digital Markets, Competition and Consumers Bill will significantly enhance the CMA's enforcement capabilities. This legislation will enable the CMA to directly impose penalties for breaches of consumer law without prolonged court procedures, thereby streamlining enforcement efforts (CMA, 2023).

Collectively, the CMA, ASA, CAP and local Trading Standards Officers provide a robust regulatory framework that ensures high standards in advertising, particularly regarding environmental claims. Adherence to these guidelines and understanding the legal implications of non-compliance are critical for all businesses operating within the UK's THE sectors.

The seven sins of greenwashing

Greenwashing can be spotted by knowing what to look for. For example, hotels may use signs encouraging water conservation without taking environmental responsibility. Vague language and buzzwords without concrete actions may reveal a company's sustainability discrepancy. A company's annual report, website and social media channels can provide further information. Misleading visuals, such as green colour palettes or images of nature without context, could be common indicators of greenwashing.

Companies may highlight minor sustainability practices to appear sustainable, masking larger adverse environmental impacts. It is important to consider the bigger picture, paying attention to their overall environmental and socio-cultural impact. Companies may boast certification labels to back up their sustainability claims. However, not all labels hold the same level of credibility. Research should be conducted about the certification programmes used by different businesses to identify those with rigorous criteria and third-party verification.

Certification is not the only solution. Some companies cannot afford the cost of certification, but that does not mean they are not sustainable. It is not enough for a certified company to merely have the required policies in place; they must create systematic change and put the policies into practice. To detect potential greenwashing, it is important to pay attention to whether companies are actively implementing sustainable practices in their operations. The TerraChoice Group's seven sins framework (2010) can be useful to the readers for initial identification and categorizing of greenwashing tactics (see Table 9.2).

From compliance to accountability: Regulation, litigation and the future of credible sustainability

The past decade has witnessed a notable increase in both public and regulatory scrutiny of sustainability claims across sectors, including THE. What was once a matter of ethical branding has rapidly evolved into a compliance issue with legal and reputational consequences. Greenwashing is no longer just a credibility risk; it is becoming a legal one. As frameworks tighten and stakeholder expectations rise, THE operators must engage more seriously with the systems of accountability that govern sustainability communication.

Fragmented rules: The challenge of regulatory incoherence

A significant hurdle facing professionals and educators in THE is navigating a fragmented regulatory landscape. Despite the growing importance of frank and honest sustainability communication, no universally accepted definition of greenwashing exists. Instead, jurisdictions have developed their own rules, enforcement mechanisms, and penalties.

Table 9.2 The seven sins of greenwashing

Sin	Action
1. Hidden trade-off	This occurs when a product is labelled 'green' based on a few attributes, overlooking other vital environmental issues. For instance, sustainably harvested paper may still contribute to greenhouse gas emissions or use harmful bleaching.
2. No proof	Environmental claims lacking easily accessible supporting information or reliable third-party certification fall into this category, such as products claiming a percentage of post-consumer recycled content without evidence.
3. Vagueness	This sin involves claims that are poorly defined or overly broad, leading to potential misunderstanding. Terms like 'all-natural' or 'non-toxic' can be misleading as they do not necessarily imply 'green'.
4. False labels	Products that mislead consumers with false suggestions or certification-like images commit this sin. An example would be a product claiming to 'fight global warming' with a certification-like image on its packaging.
5. Irrelevance	An environmental claim that may be truthful but is unimportant or unhelpful for consumers seeking environmentally preferable products. 'CFC-free' is a common example, as CFCs are banned.
6. Lesser of two evils	This sin involves a claim that may be true within the product category, but distracts the consumer from the greater environmental impacts of the category, such as organic cigarettes or fuel-efficient sport-utility vehicles
7. Fibbing	This involves environmental claims that are simply false. Common examples include products falsely claiming to be Energy Star-certified.

Adapted from TerraChoice Group, 2010

In the European Union, the proposed Green Claims Directive sets strict requirements for substantiating environmental claims with scientific evidence and third-party verification. It sits alongside other instruments such as the Corporate Sustainability Reporting Directive (CSRD) and the forthcoming Eco-design for Sustainable Products Regulation. In the UK, the Competition and Markets Authority (CMA) enforces the Green Claims Code, while in the United States, the Federal Trade Commission continues to revise its Green Guides. Canada's Competition Bureau has also increased its capacity to investigate green claims under the Competition Act. For THE operators, many of whom work across borders, serve international guests, or franchise global brands, this patchwork creates confusion and operational risk. A sustainability claim that aligns with one jurisdiction's rules may be challenged in another. Smaller operators often lack the resources to interpret or implement multiple frameworks, exposing them to unintentional non-compliance (see Table 9.3).

Table 9.3 Patchwork of greenwashing regulation

Jurisdiction	Definition of greenwashing	Enforcement body	Types of penalties	Sector focus
EU	Defined in Green Claims Directive as unsubstantiated or misleading environmental claims	European Commission / National Authorities	Fines, injunctions, reputational damage	All consumer-facing sectors, with tourism relevant
UK	Outlined in CMA Green Claims Code	Competition and Markets Authority (CMA)	Fines, mandatory corrections, market bans	All sectors including retail and travel
US	Under review in FTC Green Guides	Federal Trade Commission (FTC)	Legal action, financial penalties	Cross-sector, including advertising and digital platforms
Canada	Covered under false advertising rules in Competition Act	Competition Bureau	Fines, cease and desist orders	General consumer protection, includes tourism marketing

The legal turn: Greenwashing as a litigation risk

Alongside regulatory enforcement, litigation is emerging as a central mechanism for challenging misleading sustainability claims. This trend is visible across multiple sectors, and is increasingly relevant to THE businesses, particularly airlines, accommodation platforms and destination marketing organizations.

In 2023, more than 20 European airlines were targeted by environmental advocacy groups for making misleading claims about carbon neutrality (European Commission, .n.d.). Class-action suits have emerged in fashion, consumer goods and finance. In some cases, competitors have initiated proceedings, arguing that misleading claims distort fair competition. The legal implications of greenwashing are expanding rapidly. Civil society groups and NGOs are increasingly empowered to take legal action under new consumer protection rules and green claims directives. Similarly, courts are showing an increasing willingness to evaluate the legitimacy of corporate environmental narratives, with rulings that can lead to damages, forced retractions, and binding commitments to change communication practices.

In the context of THE, these developments are especially significant. The sector relies heavily on intangible experiences and emotional appeal, elements that lend themselves to optimistic or aspirational claims. When these claims are vague, exaggerated or unsubstantiated, they become legal liabilities. The defence that a misleading claim was unintentional or made in good faith is increasingly insufficient. What matters is whether the claim can be substantiated and whether it could reasonably mislead a consumer (Steiner et al, 2018; Feghali, Najem and Metcalfe, 2025).

The legal turn also introduces reputational risks that extend beyond the courtroom. Once targeted, companies may face prolonged media scrutiny, social media backlash and the loss of public trust. Investors are also increasingly attuned to litigation risk, incorporating it into ESG assessments and shareholder engagement strategies. In this climate, proactive legal literacy and internal accountability mechanisms are becoming not just advisable, but essential. As litigation risk rises, some companies are responding by withdrawing or minimising their public sustainability messaging. This phenomenon reflects a growing anxiety about exposure. Yet this silence can be just as damaging. Without transparent communication, there is less public accountability, less peer learning and diminished trust. For educators and professionals in THE, the challenge is to foster the skills and confidence needed to communicate sustainability truthfully, even in uncertain regulatory environments (Steiner et al, 2018; Feghali et al, 2025).

Future-proofing THE: Professional and organizational responses

Rather than retreat into silence or rely on vague aspirational language, THE organizations can take steps to embed credibility into their sustainability strategies:

- Invest in sustainability reporting frameworks aligned with recognized standards (e.g. ISO, GRI).
- Ensure claims are evidence-based, specific and independently verifiable.
- Offer staff training on greenwashing risks and communication ethics.
- Engage third-party verification where appropriate.

From an educational perspective, sustainability literacy now requires legal and regulatory awareness. Professionals need to be equipped not only to design ethical initiatives but to express them responsibly, recognizing their potential legal and reputational consequences. By embedding these capacities into curriculum design, training and leadership development, THE can shift from reactive compliance to proactive credibility.

The movement away from marketing-driven claims toward evidence-led communication is not a constraint, it is a maturity. It reflects an industry not afraid to speak honestly, act transparently and build trust in a climate of rising expectations. As regulation and litigation continue to shape the sustainability landscape, these qualities will be not only ethical strengths but strategic necessities.

REAL-WORLD EXAMPLE Ryanair: A global violation

Having explored the growing threat of greenwashing and the regulations implemented by the EU and UK to combat this deceptive practice, here is a real-world example to further illustrate the significant ethical issues and corporate culture that can lead to greenwashing violations. Ryanair's low emissions claim violation is a valid example of greenwashing (whether it was intentional or unintentional), for the ethical concerns it raises and the effects it has had on the transport industry.

Most importantly, Ryanair accepted the verdict. The ASA ruled that Ryanair's claims of being 'Europe's lowest emissions airline' were misleading. In response, Ryanair stated that it would comply with the ruling, although it reiterated its claim that its emissions per passenger kilometre are lower than those of other major European airlines.

The context

Ryanair, a well-known European airline, launched an advertising campaign positioning itself as 'Europe's lowest emissions airline' and a 'low CO_2 emissions airline'. These claims were broadly disseminated through various media channels, including press advertisements, television commercials and radio spots. The bold nature of the claims quickly drew attention and prompted complaints to the ASA, with concerns raised about potential misleading statements.

In defence of its claims, Ryanair referenced its relatively young fleet of aircraft, asserting that these newer planes were equipped with fuel-efficient engines. The airline also highlighted its high passenger load factor as a contributing factor to its lower emissions per passenger. To further substantiate their assertions, Ryanair presented data from recognized aviation organizations, Eurocontrol and Brighter Planet.

Initially, the advertisements gained approval from regulatory bodies such as Clearcast and Radiocentre, lending an air of legitimacy to the airline's environmental claims.

The violation

On closer inspection, the ASA determined that Ryanair's environmental claims were unsubstantiated and thus misleading. A significant issue identified by the ASA was the use of outdated data; Ryanair had used an airline efficiency ranking dating back to 2011, making it inappropriate for accurately assessing conditions in 2019. Additionally, the ASA took issue with Ryanair's ambiguous use of the term 'a major airline', deeming the definition unclear and unsuitable for valid comparative analysis.

The regulatory body concluded that the evidence provided by Ryanair was inadequate to support its claims. Consequently, the ASA took regulatory action by banning Ryanair's advertisements in their existing format.

Lessons learnt

This case underscores the crucial importance of accuracy in making environmental claims. It emphasizes the need for businesses to ensure that all statements are supported by current, comprehensive and relevant data. Companies should also have a thorough understanding of regulatory frameworks and actively comply with advertising standards to avoid similar pitfalls.

Moreover, businesses must recognize the significant role consumer trust plays in their brand reputation. Misleading claims not only invite regulatory action, but also risk long-term damage to consumer perception and trust. Companies are thus advised to proactively review and regularly update their claims to reflect evolving standards and data.

Takeaways

The Ryanair case study highlights several critical takeaways. Firstly, it demonstrates that advertising claims – particularly environmental ones – are rigorously scrutinized by regulatory bodies such as the ASA. Misleading advertisements are likely to be challenged and banned if found unsubstantiated.

Secondly, the relevance and currency of supporting data is crucial, as reliance on outdated or incomplete data can quickly undermine the credibility of assertions. Clearly defining comparative terms and criteria is also essential for making valid environmental comparisons.

Lastly, businesses must remain agile, adapting quickly to regulatory findings and ensuring ongoing compliance. Misleading claims, as illustrated by this case, can have lasting negative impacts on consumer perception and overall brand integrity. Ultimately, transparency, accuracy and regulatory compliance are fundamental for businesses aiming to make credible environmental claims.

Source: Hutton, 2020

REFLECTIVE QUESTIONS

1. What were the main claims made by Ryanair in their adverts?
2. Why did the ASA rule against Ryanair's claims?
3. What evidence did Ryanair provide to support its claims?
4. What lessons can be learned from this example?

This real-world example highlights the critical importance of accuracy and transparency in environmental claims. Companies must ensure that their advertising is based on current, relevant data and clearly defined terms to maintain consumer trust and comply with regulatory standards. Misleading claims not only damage reputation but also undermine efforts towards genuine environmental sustainability.

This section has looked at the role of regulatory bodies, both in the UK and the EU, in managing greenwashing. These organizations play a vital part in maintaining honesty in environmental claims, ensuring that all stakeholders can trust the 'green' credentials that companies present. In the next section, the focus moves away from the complicated and broad landscape of regulations and their enforcers to the ground level, in the world of the THE sector.

Greenwashing in THE operations

In this section, other real-world examples of greenwashing within the THE sector are explored. It will be seen how companies, in their quest to appear environmentally friendly, can sometimes overstate their green initiatives. These examples will not only make the concept of greenwashing more tangible but also underline the importance of the regulatory frameworks just discussed. The examples provided below merely illustrate the possible risks associated with greenwashing. They do not, unless explicitly mentioned, refer to any specific organizations.

Animal experiences

Animal experiences are a major draw for tourists; however, not all these experiences are ethical or contribute to wildlife conservation. Some of the so-called 'sanctuaries' and 'eco-tours' prioritize entertainment over animal welfare, and engage in harmful practices such as forced breeding, early separation from mothers, and even physical abuse. For example, elephant bathing is a popular tourist activity that often involves coercive training methods and disrupts the animals' natural habitats. Similarly, unethical practices can be seen in elephant riding and dolphin swimming experiences. Red flags indicating unethical practices include animals performing unnatural behaviours. Authentic sanctuaries or eco-tours prioritize the welfare of the animals, avoiding direct interaction such as petting or feeding.

To avoid supporting such unethical practices, thorough research is crucial. Seek guidance from credible organizations like World Animal Protection and Sustainable Travel International, as well as looking for certifications from reputable programmes such as the World Cetacean Alliance for whale and dolphin-watching operators.

Reading reviews and examining photos can also provide insight into whether tourists are allowed to touch the animals and the overall experience. Tourists should not hesitate to ask providers about the animals' origins and care. This vigilance can help ensure that tourists' choices support genuine sustainability and animal welfare.

Authentic cultural native experiences

Tourists are often attracted by seemingly 'authentic native' cultural experiences that are, on many occasions, commercialized attractions tailored to tourist expectations. Examples include shamanic rituals, which commodify and distort local cultures for entertainment. Such practices dilute the significance of local cultures, can perpetuate harmful stereotypes, and lead to divisions within vulnerable communities. Under the guise of cultural immersion, participation in inauthentic experiences can undermine local culture and disenfranchise communities without bringing economic benefits. Non-commodified experiences should be managed by local communities in conjunction with THE businesses to ensure benefits are equally benefitting the local economy.

An example of the dilution of local culture

The indigenous population of New Zealand are the Māori. Before visitors can enter a communal or sacred place that serves as a venue for social and cultural activities, called a marae, an opening welcome ceremony the 'pōwhiri' is performed.

When this ceremony is performed for the entertainment of tourists altered to cater to tourists' expectations, or simplified for easier understanding, the authenticity of the experience may be compromised. This can lead to a gradual loss of the original cultural significance and depth, making it harder for future generations of Māori to fully understand and appreciate their heritage.

Carbon neutrality and climate consciousness claims

THE companies may often aim for 'carbon neutrality' to demonstrate their commitment to combating climate change, as seen in the real-world example earlier. The United Nations defines carbon emissions as the release of greenhouse gases, including carbon dioxide (CO_2), into the atmosphere over a specified area and period of time (UN, .n.d. b). These emissions primarily result from human activities such as burning fossil fuels, deforestation and industrial processes.

However, ambiguous inaccurate claims relying on incorrect carbon footprint calculations or low-quality offsetting initiatives are misleading. Airlines and cruise liners have faced criticism for their claims on the environmental benefits of their sustainability efforts, including the use of vague terms like '100 per cent Green' or

'CO_2Zero'. They have also been criticized for promoting the use of Sustainable Aviation Fuel (SAF) without fully disclosing what percentage of SAF makes up their total fuel usage.

Consumers may assume all emissions have been eliminated when a product is marketed as carbon neutral, leading to accusations of greenwashing when they discover that the company did produce emissions but neutralized them via carbon offsets. According to the UN (.n.d. c), carbon offset refers to the reduction, avoidance or removal of greenhouse gas emissions to compensate for emissions produced elsewhere. This is typically achieved through purchasing carbon credits from projects that reduce, avoid or remove emissions, such as renewable energy projects, reforestation or energy efficiency improvements. While carbon offsets play a crucial role in climate action, they should supplement rather than replace emissions reduction tactics. Companies risk entering the murky water of greenwashing when relying solely on carbon offsets without a decarbonization plan and active steps to reduce their avoidable emissions.

Consumer participation

THE sustainability programmes often rely hugely on customer participation, shifting the burden of action onto guests. This can enhance a company's green image, but without consistent corporate action, it becomes greenwashing.

Restaurants, for example, may encourage customers to finish meals to reduce waste, but without internal changes to reduce food waste, this prompts overeating. Effective waste reduction requires appropriate practices in inventory management, menu planning and portion control. Tourists can identify greenwashing policies that place sustainability responsibility on them, rather than a mutual effort with the business, such as signs encouraging meal completion. Tour operators and airlines often ask travellers to offset their own carbon emissions, presenting it as a corporate initiative. However, these programmes often have low participation and minimal impact. THE businesses genuinely committed to reducing emissions should include offsets in every trip and strive to decarbonize and reduce avoidable emissions.

Environmentally friendly hospitality operations

Hotels and lodges claiming to be 'eco-friendly' or 'sustainable' by adopting a natural aesthetic and promoting nature-based activities might create an illusion of environmental responsibility. However, their practices could contradict their claims, by disrupting fragile ecosystems, consuming excessive resources, generating waste and contributing to deforestation and overconsumption.

For example, off-the-grid hotels may rely on diesel generators to provide consistent energy flow for power, contributing to air pollution and carbon emissions. Off-the-grid refers to living without reliance on public utilities, such as electricity, water, and gas, often by using alternative energy sources like solar power.

Hotels in water-scarce areas may maintain large swimming pools, exacerbating local water scarcity. These practices can highlight the presence of greenwashing in the sector, where the reality of operations does not match the eco-friendly image presented to the public.

Recycling practices vs recycling facilities

THE businesses may advocate the use of compostable or recyclable items as a strategy to reduce waste. However, in some destinations, the necessary local authorities recycling facilities or industrial composting capabilities to process this waste effectively are lacking or not working to their full potential. Therefore, the fate of recyclables from many hotels is that they end up in a landfill. While many hotels may have recycling programmes in place, if they do not have the facilities for sustainable waste removal, this is an example of greenwashing. The absence of a robust recycling infrastructure results in recyclable items, discarded by visitors into recycling bins, ending up in landfills.

In such scenarios, the use of recyclable or compostable single-use items makes little difference to the overall waste problem. It would be more beneficial to opt for reusable items whenever feasible to minimize the production footprint and waste generation. THE businesses that prioritize convenience over sustainability may exaggerate the role of these products in waste reduction, resulting in greenwashing.

Sustainability signage

Sustainability signage can mislead guests into believing they are contributing to eco-friendly initiatives. Hotels may profess to uphold sustainable practices, such as encouraging guests to reuse towels and linens; however, they could contradict their policies by replacing these items daily, regardless of the guests' preferences. A similar discrepancy could be observed with energy consumption. Hotels may display signs urging guests to turn off the air conditioning when they leave the room; however, if this practice is not followed up by staff, then resources are wasted and a false impression created of the hotel's sustainability commitment.

These inconsistencies between stated policies and actual practices are often unintentional, stemming from different reasons including a lack of staff training and standardized procedures. However, regardless of the intent, such practices fall under the umbrella of greenwashing. In essence, while such operations may appear to be eco-friendly on the surface, their behind-the-scenes operations can reveal a different reality. This discrepancy between their outward image and their actual practices is a classic example of greenwashing.

Having explored some of the greenwashing practices in the THE sector, the next section will provide the readers with tools to identify instances of greenwashing and some recommendations to avoid greenwashing practices.

Avoiding instances of greenwashing

Actionable strategies that THE companies can adopt to prevent greenwashing emphasize impactful sustainability initiatives and the importance of maintaining authenticity in their claims.

Increase overall awareness

To prevent instances of greenwashing, THE companies should prioritize professional development for themselves and their staff (as discussed in Chapter 3). Understanding the principles of sustainable travel and staying informed about industry guidelines and best practices is crucial. Organizations like Sustainable Travel International offer free educational resources and training programmes.

Additionally, THE could consider leveraging guidance from industry bodies such as the Global Sustainable Tourism Council (this provides guidelines and criteria for sustainable tourism practices, including animal welfare standards), Green Fins (this provides internationally recognized environmental standards for marine tourism), and Planterra (whose aims to ensure that tourism benefits local people economically, culturally and environmentally). By taking these steps, companies can actively avoid greenwashing and contribute to genuine sustainability efforts.

Create an overarching policy

While many tourism companies emphasize environmental sustainability, true sustainability extends beyond the environment. Santos, Coelho and Marques (2023) recommend that a comprehensive sustainability policy should address the four pillars of sustainability: environment, social, economic and cultural (as discussed in Chapter 2). When crafting such a policy, THE should consider its impact across all four pillars by:

- **relating**, to demonstrate 'caring for others and the world', recognizing the environment as a stakeholder and involving local or indigenous community knowledge and experiences (Inner Development Goals, .n.d.)

- **collaborating,** to demonstrate consideration of individuals and groups who are connected with and affected by the THE's impact upon the environment (Inner Development Goals, .n.d.)
- **acting,** to demonstrate a commitment to enabling change by taking decisive action to develop innovative and ethical operations (Inner Development Goals, .n.d.)

Consistency: Claim vs practice

To avoid greenwashing, THE should ensure that words align with tangible actions. A sustainability plan should include the professional development of staff on the plan's impact and implementation. This will encourage transparency, which would prevent accusations of greenwashing by fostering genuine commitment. Understanding the 'why' behind each initiative in the sustainability plan can motivate employees to adopt new practices more readily. Employees serve as the face of THE businesses. By helping staff to understand the importance of sustainable practices, they are empowered to act as stewards of sustainability and effectively communicate the THE's business commitment to customers. Honest conversations with guests about the business's sustainability efforts lend transparency and help avoid accusations of greenwashing.

Avoid vague language

When promoting sustainability, THE companies must exercise caution with terms such as 'green', 'environmentally friendly' or 'carbon neutral'. To avoid being misleading, THE companies should thoroughly understand the criteria and regulations behind such claims. As previously seen in this chapter, the EU's proposal aims to establish strict standards for environmental marketing. Ideally, THE companies should avoid vague terms altogether, opting for specific statements such as 'our property operates on 90 per cent renewable energy' to convey genuine commitment.

Provide clear evidence

THE companies must provide data and other type of evidence to substantiate sustainability claims. Visual aids, such as photos or videos, can make claims tangible. For instance, if a cruise ship claims water reduction, it should be able to demonstrate the percentage decrease through tracking mechanisms. For example, they could accurately report whether the water reduction was the result of the installation of low-flow taps or the implementation of greywater reuse. Real-life stories can enhance credibility.

Building trust

Building consumer trust links back to transparency, by being both honest and transparent. For any THE business, providing one small example of a sustainability initiative is better than making exaggerated claims that do not stand up to scrutiny. For example, when making carbon neutral claims, THE companies should explicitly define their approach, measured emission sources, methodology and total carbon footprint.

Conduct research

As part of a rigorous sustainability plan THE companies should make certain that research is a component of the plan. Conducting research into the legal requirement, policies and procedures will ensure an informed approach for each planned sustainability initiative. This could include collaborating with experts such as Sustainable Travel International or charitable organization such as WRAP. Additionally, when carrying out sustainability projects, THE companies should verify them against rigorous standards like Plan Vivo, Verified Carbon Standard or the Gold Standard and consider whether incorporating offsets into the company's budget is an ethical option.

Short termism

To avoid greenwashing THE companies should refrain from short term practices such as offsetting carbon emissions. Offsetting carbon emission does not change internal practices. THE companies should demonstrate clear decarbonization strategies and reconfigure operations to reduce and eventually eliminate emissions at their source. Thus, decarbonization efforts can then ensure genuine carbon footprint reduction while supporting broader environmental goals.

The danger of greenhushing

While scrutinizing sustainability claims is crucial, excessive scepticism can lead to 'greenhushing.' Greenhushing occurs when companies stay silent about their sustainability efforts, fearing accusations of greenwashing. However, this stifles progress toward industry-wide sustainability initiatives. Greenhushing suggests that sustainability lacks reward and invites criticism. There needs to be a balance between accountability and the genuine efforts toward becoming more sustainable. Greenhushing affects one in four companies (Letzing, 2022), and it is seen to hamper transparency, making it harder to make the informed decisions needed in the THE industry as a whole.

> **INDUSTRY VOICE: SUE WILLIAMS**
>
> Sue Williams, an expert in sustainability and a former GM, says, 'The tolerance for vague "green" claims is disappearing, especially with stricter regulations like the EU's Green Claims Directive demanding verifiable proof. Genuine sustainability isn't just about good intentions; it requires rigour, transparency and data. We found that using science-based certifications and audits, like EarthCheck, was crucial not just for measuring our actual impact, but for demonstrating authentic commitment and building trust. Businesses need an inquiring mind – investigate claims, demand evidence, and be prepared to prove your own. Avoiding greenwashing isn't just about compliance; it's fundamental to credibility and long-term resilience in this industry.'

Conclusion

This chapter has explored the critical issue of greenwashing in the THE industry, examining its definitions, motivations and negative consequences. Key regulations in the EU and UK aimed at ensuring truthful communication were reviewed, alongside practical tools, such as the seven sins framework and illustrative examples, for identifying misleading environmental claims. Emphasis was placed on actionable strategies for businesses to avoid greenwashing by prioritizing transparency, substantiating claims with clear evidence, and ensuring consistency between communication and practice.

Recognizing and rejecting greenwashing is crucial not only for compliance, but for building stakeholder trust and ensuring the credibility of the THE sector's contribution to genuine sustainability. The following student activity offers a platform to apply these insights and hone critical evaluation skills.

> **ACTIVITY 1: ANALYSING GREENWASHING CLAIMS**
>
> This task is independent work and should take approximately 30 minutes.
>
> *Purpose*
>
> Sharpen your ability to quickly analyze potential greenwashing in the THE industry and practice forming arguments – key skills for ethical practice and debate. Remember, greenwashing involves potentially misleading claims about environmental practices.

Instructions

1. **Review the scenario (5 mins)**

   Based on common issues in the THE industry (drawing on examples like those discussed in your chapter), read the short scenario below:

   A large hotel promotes its new 'Eco-Stay Initiative'. Key messages include:

   o 'Join us in saving the planet – reuse your towels!' (Signs in bathrooms).

   o 'We're reducing our carbon footprint with energy-efficient lighting in all guest rooms.'

   o 'Enjoy our locally-sourced breakfast options – kinder to the environment!'

   Note: The hotel uses single-use plastics extensively, sources most energy from fossil fuels, and claim 'locally sourced' food

2. **Identify potential issues (5 mins)**

   Based on the seven sins framework and general concepts from your chapter, identify any potential greenwashing issues you spot in the hotel's claims (e.g. hidden trade-offs, vagueness, lesser of two evils).

3. **Formulate arguments (15 mins)**

   Consider the statement: 'The hotel's Eco-Stay Initiative claims are acceptable marketing practices.' Focus on using reasoning based directly on the scenario and the greenwashing concepts you have learned:

   o Write two brief arguments supporting this statement. (Why might someone argue these claims are acceptable or at least understandable in a business context?)

   o Write two brief arguments opposing this statement. (Why are these claims potentially misleading or unethical greenwashing?)

4. **Quick reflection (5 mins)**

   o Which side (supporting or opposing) did you find easier to argue for? Why?

   o Briefly state one key ethical consideration this scenario highlights for THE businesses.

KEY TAKEAWAYS

- Understanding greenwashing:
  o Greenwashing involves deceptive marketing practices where companies falsely promote their products or services as environmentally friendly.
  o It is crucial to recognize the history and definition of greenwashing to understand its impact on the THE industry.

- Regulatory frameworks:
    - The chapter examines the regulatory frameworks in the EU and UK that aim to combat greenwashing.
    - Awareness of these regulations helps in identifying and addressing greenwashing practices.
- Deceptive tactics:
    - Companies often use vague claims or misleading net-zero commitments to appear more sustainable than they are.
    - Recognizing these tactics is essential for promoting genuine sustainability.
- Real-life examples and legal precedents:
    - By studying real-life instances and legal precedents, students can identify patterns and trends in greenwashing.
    - This knowledge helps in understanding the human factor in interventions and the broader impact on the industry.
- Tools and strategies:
    - The seven sins framework is introduced as a tool to identify deceptive claims.
    - Students, lecturers and professionals are encouraged to innovate and implement effective strategies to enhance greenwashing awareness and combat it.

REFLECTIVE QUESTIONS

1. What is the definition of greenwashing?
2. How does the chapter clarify existing regulations in EU and UK?
3. Describe the instances of greenwashing in THE.
4. What can THE operators do to address instances of greenwashing?

Further resources

ASA (2024) Ruling. Advertising Standards Authority, www.asa.org.uk/codes-and-rulings/rulings.html (archived at https://perma.cc/GZK6-DAS9)

EU (2024) Stopping greenwashing: how the EU regulates green claims. Topics European Parliament, www.europarl.europa.eu/topics/en/article/20240111STO16722/stopping-greenwashing-how-the-eu-regulates-green-claims (archived at https://perma.cc/UQ25-78C7)

Wang, D, Walker, T and Barabanov, S (2020) A psychological approach to regaining consumer trust after greenwashing: The case of Chinese green consumers, *Journal of Consumer Marketing*, **37** (6), pp 593–603

References

Baude, W and Sachs, S S (2016) The law of interpretation, *Harvard Law Review*, **130**, p 1079, https://harvardlawreview.org/print/vol-130/the-law-of-interpretation/ (archived at https://perma.cc/8DWM-268D)

Choudhury, R R, Islam, A F and Sujauddin, M (2024) More than just a business ploy? Greenwashing as a barrier to circular economy and sustainable development: A case study-based critical review, *Circular Economy and Sustainability*, **4** (1), pp 233–66

CMA (2021a) Global sweep finds 40% of firms' green claims could be misleading, GOV.UK, 28 January, www.gov.uk/government/news/global-sweep-finds-40-of-firms-green-claims-could-be-misleading (archived at https://perma.cc/8DWM-268D)

CMA (2021b) CMA guidance on environmental claims on goods and services, GOV.UK, 20 September, www.gov.uk/government/publications/green-claims-code-making-environmental-claims/environmental-claims-on-goods-and-services (archived at https://perma.cc/XSC9-9N2L)

CMA (2023) New bill to stamp out unfair practices and promote competition in digital markets. GOV.UK, 25 April, www.gov.uk/government/news/new-bill-to-stamp-out-unfair-practices-and-promote-competition-in-digital-markets (archived at https://perma.cc/8NV9-VVS3)

Committee of Advertising Practice (2023) Advertising Guidance: misleading claims and social responsibility in advertising, 23 June, ASA, www.asa.org.uk/resource/advertising-guidance-misleading-environmental-claims-and-social-responsibility.html (archived at https://perma.cc/JZP4-BN9R)

de Freitas Netto, S V, Sobral, M F F, Ribeiro, A R B and Soares, G R D L (2020) Concepts and forms of greenwashing: A systematic review, *Environmental Sciences Europe*, **32**, pp 1–12

European Commission (2023) Proposal for a DIRECTIVE OF THE EUROPEAN PARLIAMENT AND OF THE COUNCIL on substantiation and communication of explicit environmental claims (Green Claims Directive), 22 March, https://eur-lex.europa.eu/legal-content/EN/TXT/PDF/?uri=CELEX:52023PC0166 (archived at https://perma.cc/79EK-YR66)

European Commission (.n.d.) Green Claims, https://environment.ec.europa.eu/topics/circular-economy/green-claims_en (archived at https://perma.cc/G54D-4JTM)

European Securities and Market Authority (2024) Final Report on Greenwashing, 4 June, www.esma.europa.eu/sites/default/files/2024-06/ESMA36-287652198-2699_Final_Report_on_Greenwashing.pdf#page=78.16 (archived at https://perma.cc/9TT7-F5BM)

European Union Agency for Fundamental Rights (2024) Enforcing consumer rights to combat greenwashing, 7 March, https://fra.europa.eu/en/publication/2024/enforcing-consumer-rights-combat-greenwashing (archived at https://perma.cc/Z9RT-YUWD)

Feghali, K, Najem, R and Metcalfe, B (2025) Greenwashing in the era of sustainability: A systematic literature review, *Corporate Governance and Sustainability Review*, **9** (1), pp 18–31, https://doi.org/10.22495/cgsrv9i1p2 (archived at https://perma.cc/C6YC-PHE2)

Hutton, R (2020) Ryanair rapped over low emissions claims, BBC News, 5 February, www.bbc.co.uk/news/business-51372780 (archived at https://perma.cc/KVF4-JPYU)

Inner Development Goals (.n.d.) Inner Development Goals Framework, https://innerdevelopmentgoals.org/framework/ (archived at https://perma.cc/D3WP-9EK9)

Letzing, J (2022) What is 'greenhushing' and is it really a cause for concern?, World Economic Forum, 18 November, www.weforum.org/agenda/2022/11/what-is-greenhushing-and-is-it-really-a-cause-for-concern/ (archived at https://perma.cc/8YCF-TNH5)

Orange, E and Cohen, A M (2010) From eco-friendly to eco-intelligent, *The Futurist*, **44** (5), p 28

Santos, C, Coelho, A and Marques, A (2023) A systematic literature review on greenwashing and its relationship to stakeholders: State of art and future research agenda, *Management Review Quarterly*, pp 1–25

Steiner, G, Geissler, B, Schreder, G and Zenk, L (2018) Living sustainability, or merely pretending? From explicit self-report measures to implicit cognition, *Sustainability Science*, **13** (4), pp 1001–15, https://doi.org/10.1007/s11625-018-0561-6 (archived at https://perma.cc/8FJQ-49TV)

TerraChoice (2010) The sins of greenwashing: home and family edition, http://sinsofgreenwashing.org/fndings/the-seven-sins/ (archived at https://perma.cc/PYK7-MA34)

UN (.n.d. a) Greenwashing – the deceptive tactics behind environmental claims, www.un.org/en/climatechange/science/climate-issues/greenwashing (archived at https://perma.cc/V4FS-WUXP)

UN (.n.d. b) What is climate change? www.un.org/en/climatechange/what-is-climate-change (archived at https://perma.cc/8JWP-UNTK)

UN (.n.d. c) United Nations Carbon Offset Platform, https://unfccc.int/climate-action/united-nations-carbon-offset-platform (archived at https://perma.cc/UG3V-JP7C)

10 | Conclusion

Sustainability as practice in THE

This concluding chapter brings together the key ideas, values and practices explored throughout the book. Its purpose is not to summarise, but to extend the conversation. Sustainability has been approached throughout as a professional responsibility, supporting your transition from learning into practice.

The structure of the book has reflected a broader understanding of sustainability learning. We have explored ways of thinking, ways of doing, and ways of being. These modes of engagement support the development of your competencies for sustainability as you prepare to enter professional roles in THE.

These competencies align with global frameworks such as the Sustainable Development Goals (SDGs) and are reinforced by the work of agencies such as UN Tourism. These are not templates to be followed, but evolving structures and influences that guide decisions and shape professional environments. Understanding how these elements operate in practice will be part of your role as a future professional.

As you enter THE sectors, you will find that sustainability already influences how organizations operate, how strategies are evaluated, and how value is defined. You will be expected to work in partnership with stakeholders across sectors, roles and geographies. We have presented real-world examples throughout this book to illustrate how SDG 17 supports this collaboration, guiding environmental, social and governance (ESG) practices and shaping how corporate social responsibility (CSR) is interpreted in context.

In this concluding chapter, we do not revisit whether sustainability matters. That case has been established. Sustainability is now an imperative of professionalism in THE. What remains is how you will take this work forward as an emerging practitioner. This chapter draws together the key learning from the book and positions you in your professional capacity as an agent for change within it.

Framing the professional landscape

Sustainability in THE emerges through the systems, policies and infrastructures that govern how services are delivered, how resources are managed, and how priorities are set. Professionals entering the field encounter these conditions already in motion, some supportive of sustainable practice, others more resistant to change. This section considers how professionals can work within these systems while remaining attentive to the possibilities for influence, adaptation and ethical engagement.

Working with sustainability frameworks in context

Global frameworks such as the Sustainable Development Goals (SDGs) and the Measuring Sustainable Tourism (SF-MST) framework provide essential reference points for sustainability in THE. Their influence arises from how professionals interpret and enact them within organizational, cultural and operational settings. These frameworks offer a shared vocabulary through which intentions, priorities and performance standards are articulated, negotiated and applied.

In practice, these frameworks take shape through specific tools and metrics such as carbon audits, climate vulnerability assessments and social equity indicators. Used with critical intent, these instruments guide strategic reflection and drive improvements in service design, delivery, and evaluation. They reveal disconnects between institutional sustainability goals and operational realities, highlighting areas that demand professional leadership and systemic adjustment.

To drive meaningful change, sustainability frameworks must be fully integrated into decision-making systems. When procurement strategies, performance evaluations and partnership models align with sustainability goals, they reshape how value is defined and how accountability is practiced. In these conditions, sustainability functions as a core principle of organizational operations.

Professionals lead this integration by applying frameworks with strategic focus. Some coordinate them across departments to align goals with metrics. Others tailor them to local contexts, ensuring relevance and responsiveness to stakeholder needs. Examples throughout this book demonstrate that sustainability frameworks generate their greatest value through professional judgement, contextual awareness and informed application.

Working with frameworks in this way requires active engagement with the assumptions they carry, the behaviours they influence and the systems they help to shape. These tools function most effectively when treated as dynamic instruments of sustainability practice, capable of advancing transformation when used with insight and intention.

Engaging frameworks critically and constructively expresses a commitment to sustainability as a professional standard. For emerging professionals, the responsibility is to interpret, adapt and apply these frameworks in ways that support context-responsive and ethically-grounded practice.

Operationalizing circularity in THE

Circular economy models restructure how value is created and sustained in THE. These models reduce waste, extend resource life, and support ecological renewal. In THE circularity informs procurement practices, energy systems, food supply chains and service design. It replaces the extractive logic of take–make–dispose with an emphasis on keeping materials in use through recovery and redesign.

The 8Rs framework introduced earlier—Refuse, Rethink, Reduce, Reuse, Repair, Refurbish, Remanufacture and Recycle, provides a practical structure for interpreting circularity in professional contexts. These strategies take on greater significance when embedded across operations. Applied systematically, they alter how resources are sourced, how services are delivered, and how value is maintained throughout the business cycle.

Many professionals will enter organizations still governed by linear practices or only beginning to adopt circular models. In these environments, circular initiatives may emerge gradually through small-scale waste reduction, sourcing pilots or staff-led adjustments. Progress in such settings depends on more than systems or tools. Circularity becomes embedded when leadership prioritizes it and when sustainability is treated as a strategic imperative, not a technical supplement.

Examples across THE show circular approaches in action from food redistribution and adaptive menus to shared infrastructure and low-carbon accommodation. These shifts do more than improve efficiency. They signal a broader redefinition of what constitutes responsible and resilient practice.

Circular tools require professionals to coordinate across functions, navigate supply networks, and identify where decisions intersect. A change in waste processes, for instance, might expose weaknesses in procurement or suggest missed opportunities for regional collaboration. Effective professionals read these patterns and act with awareness of the broader systems at play.

Applied deliberately, circular principles help professionals identify leverage points for change. The aim is to build capacity for systems-level adaptation. This includes refining existing processes, forging new connections and aligning decisions with the long-term viability of the enterprise. For professionals entering THE circularity is not a technical add-on, but rather an approach carried into practice within complex, imperfect systems. It is a mindset that enables high-level goals to be translated into consistent, context-sensitive decisions.

Practising sustainability in complex and imperfect systems

Sustainability in THE takes shape through systems that are dynamic and often inconsistent. Financial models, regulatory frameworks and workplace cultures all influence how sustainability is defined and delivered. For professionals entering the field, learning to work within these systems is essential to acting with purpose and insight.

The pressures that drive sustainability such as climate disruption, resource constraints and changing social expectations are reshaping how services are planned and managed. These shifts are also influencing the policy landscape. Initiatives like the European Green Deal, the Corporate Sustainability Reporting Directive (CSRD), and the Sustainable Development Goals (SDGs) offer frameworks for coordinated action. Their relevance, however, depends on how they are translated into practice. Professionals contribute to this work by interpreting broad goals in relation to the priorities and realities of their own settings.

In applying these goals, professionals often face decisions shaped by competing demands. A shift in procurement may reduce emissions but increase costs. A conservation measure may protect biodiversity while restricting access. These trade-offs are common, and they require interpretation rather than automatic compliance. They call for professional judgement that balances intent with outcome, and ideals with the conditions in which choices are made.

That judgement is shaped by the context of work. Organizational culture, leadership priorities and available resources affect how sustainability can be pursued day to day. While some professionals new to THE will join organizations with clear direction and support, others will encounter fragmented priorities or limited capacity. In both situations, understanding the work environment becomes central to acting effectively.

This capacity to read context is what enables professional agency. Influence is exercised through the ability to recognize what is possible, where change can occur, and how alternatives can be proposed. Integrity is maintained through careful observation and timely response, not by waiting for ideal conditions.

Professionals build their capacity to act through deliberate practice. The SDG learning framework identifies key competencies that support this process, such as the ability to analyse complexity or respond constructively to evolving conditions. These sustainability competencies help professionals exercise judgement, recognize opportunity and stay focused when navigating uncertainty.

Across these conditions, sustainability becomes real through consistent, considered practice. THE professionals who engage with complexity contribute to outcomes that are responsive, responsible and credible over time.

Sustaining place: Culture, heritage and responsibility

Professionals working in THE operate within systems that influence not only economic outcomes but also the cultural and ecological integrity of place. Visitor economies are shaped by social, environmental and political forces that determine whose stories are told, whose knowledge is centred, and whose interests are advanced. For those entering the sector, engaging with sustainability requires close attention to how professional decisions affect the wellbeing of place.

These broader systems influence not just how THE operates, but how it contributes to the resilience or vulnerability of the destination it serves. Tourism plays a dual role in many destinations. It contributes to environmental and cultural pressures, yet it also holds potential to support recovery and adaptation. When ecosystems are disrupted, tourism operations can either intensify local vulnerabilities or help rebuild community strength. The outcome depends on how professionals respond to local priorities, how organizations structure partnerships, and how decisions are made about whose needs guide strategy. Initiatives that engage local networks and build on cultural practices are more likely to sustain both people and place.

Beyond infrastructure and service delivery, sustainability also depends on how professionals engage with local cultural meaning and identity. Culture and heritage are central to how communities define their wellbeing. Local memory and language are not marketing assets they are forms of identity rooted in context. When professionals design experiences without recognizing these roots, they risk flattening meaning and weakening the distinctiveness of a destination.

This relationship between identity and place is also reflected in how the physical environment is understood and valued. Built and natural environments carry significance beyond aesthetics or utility. Landscapes embody historical relationships, ecological systems or spiritual meaning. In some contexts, they anchor traditions and connect communities across generations. When professionals treat the environment as a stakeholder, they position sustainability as a relationship to be maintained not a problem to be managed.

Recognizing these layers of meaning can clarifies the ethical significance of decisions that affect place and community. In settings shaped by overtourism or commercial pressure, professionals are often required to make decisions that challenge operational expectations. Limiting visitor numbers, revising programme content or declining certain initiatives may be unpopular but these choices are often necessary to uphold cultural continuity and respect community-led priorities.

In such settings, the strength of sustainability commitment is demonstrated through the way professionals behave in response to local conditions. Sustainable practice in this context calls for more than awareness. It calls for active engagement with both the visible conditions of place and the deeper historical forces that shape them. Sustainable practice in this context involves more than observation. Professionals must listen to community voices, understand the legacies that influence present conditions, and contribute to futures defined by respect and shared purpose. Working in this way affirms dignity and responsibility. It also signals a commitment to care that responds directly to the complexities and significance of place.

Interpreting practice in context

Professional sustainability practice is shaped as much by interpretation as by instruction. Mindset, governance and futures thinking are not abstract concepts, they are enacted through decisions made within imperfect conditions and alongside competing demands. This section explores how professionals make sense of sustainability in the contexts they inherit, and how they use judgement to align actions with values. These are the moments where sustainability is most often tested and where its credibility is built.

A flexible mindset for navigating sustainability

Sustainability in THE calls for professionals who apply knowledge with ethical awareness and contextual judgement. Their effectiveness is shaped not only by individual expertise but by the environments in which they operate.

A flexible mindset is essential to navigating this complexity. Professionals interpret dynamic conditions, shifts in climate, markets, or community expectations, and respond with focus and intent. Flexibility supports adaptive decision-making grounded in professional responsibility. Organizations that prioritize this responziveness build the capacity to adjust operations as conditions evolve. Flexibility becomes a defining feature of professional culture, reinforced through expectations and demonstrated in practice.

This responsiveness is underpinned by specific competencies. Systems thinking, ethical reasoning and contextual awareness express the applied judgement needed in sustainability practice. This perspective aligns with UNESCO's SDG competency framework, which highlights critical thinking (cognitive), self-awareness (socio-emotional), and collaborative action (behavioural) as to future-oriented professional learning. These competencies enable professionals to act decisively within fluid and often contradictory settings.

Competence develops through applied experience. Professionals strengthen their capabilities by working through real constraints, interpreting conflicting demands and adjusting their reasoning under pressure. Learning is supported when organizations create conditions for reflection and experimentation, allowing insight to emerge through use rather than instruction alone.

Throughout this book, examples have shown that sustainability becomes embedded through collaborative action. Professionals co-create understanding with others, listen to context-specific knowledge, and surface assumptions that might otherwise remain unchallenged. These practices refine professional judgement and ensure that ethical considerations remain visible and active within decision-making.

This defines ethical stewardship in practice. THE professionals act with foresight, apply their values through grounded decisions, and remain attentive to the effects of their actions. Stewardship becomes embedded when sustainability is treated as an everyday standard of practice rather than a separate issue to be addressed in isolation.

Futures thinking and strategic judgement

This book has approached sustainability as a process shaped by evolving systems. Futures thinking has been made implicit through emphasis on adaptation, context-sensitive decision-making, and on the need to respond when conditions change. This section brings those elements into clearer focus, affirming futures thinking as a core competency of professional judgement.

Futures thinking refers to the structured exploration of alternative futures to inform decisions in the present. It involves analysing change and assessing long-term implications. Within THE, it enables professionals to respond more effectively to instability and shifting priorities, conditions that increasingly shape the operating environment.

This capacity to work with uncertainty is now central to professional competence. Understanding sustainability as a future-oriented process requires acknowledging that change is ongoing and rarely predictable. Professionals entering the sector must be prepared to work in conditions that do not follow familiar patterns. This means learning to interpret uncertainty while maintaining clarity of purpose.

In this context, futures thinking is not speculative forecasting. It is a structured form of analysis linking current decisions to emerging conditions and long-term outcomes. By applying futures thinking, professionals maintain the ability to act with purpose and adapt strategy as new information emerges.

Futures literacy, as outlined by UNESCO, strengthens professional judgement by encouraging reflection on assumptions, exposure to alternative scenarios, and openness to revised strategies. In THE, it enhances operations by furthering responsiveness and supporting decisions that remain aligned with evolving systems.

Developing this capability involves intellectual and ethical effort. It requires a willingness to challenge assumptions and examine underlying systems. These practices move sustainability from technical response to strategic judgement.

For new professionals in THE futures thinking is not an optional perspective but a core competence of practice. It supports the capacity to make considered decisions when conditions are shifting, and outcomes are not guaranteed. This is the context in which sustainability work increasingly takes place. Responding with clarity, consistency and relevance is not a matter of prediction, but of professional judgement applied over time.

Shaping governance through professional practice

Acting with foresight requires more than individual awareness. It depends on the structures through which decisions are made, justified and shared. Professionals do not operate outside these systems. They contribute to shaping them. Governance, in this context, is an ongoing process of negotiation, in which values are contested, priorities are set, and outcomes are defined. This section examines how professionals participate in governance by influencing how sustainability is interpreted and operationalised through everyday decisions.

Professionals entering THE sectors work within policy environments that define expectations, but do not determine outcomes. Frameworks such as the Sustainable Development Goals (SDGs), the European Green Deal and the Corporate Sustainability Reporting Directive (CSRD) create reference points for action. Their significance, however, is shaped through implementation. Professionals play a role in how strategies are interpreted and connected to the specific demands of place and culture.

Participating in governance means recognizing how decisions are structured and where professional judgement can shape outcomes. This includes aligning procurement with sustainability principles, contributing to transparent reporting, and supporting participatory destination planning. These are not administrative tasks. They are practices through which values become embedded in systems.

Effective governance also requires responsiveness to change. Sustainability frameworks must evolve with shifting environmental and social conditions. Professionals help maintain this relevance by interpreting policy in context, identifying gaps in application, and supporting strategies that reflect both ecological priorities and social accountability.

Shared responsibility in sustainability governance is enacted through the coordination of decisions across organizations and stakeholder groups with overlapping responsibilities. While early-career professionals may not design policies, they participate in how those policies are interpreted, enacted and communicated. This includes identifying points of alignment, recognzing when certain voices are excluded, and supporting efforts to broaden participation. Developing this capacity requires attention to process as well as outcomes, and a clear understanding of how decisions affect equity and accountability on the ground.

Governance is not a neutral backdrop to professional activity. It is a system that both shapes, and is shaped by, everyday decisions. For new professionals, influence may begin with small interventions such as the framing of reports, or the structuring of meetings, or the management of stakeholder dialogue. These forms of engagement may be indirect, but they contribute to how sustainability priorities are maintained and negotiated in practice.

Moving forward

Maintaining sustainability work requires more than policy alignment or technical expertise. It depends on a continued willingness to learn, to reflect, and to invest in one's own development over time. Education and wellbeing are not peripheral concerns; they support the endurance and integrity of professional practice. This section considers how professionals can remain engaged in the long term, be responsive to change, accountable in decision-making, and committed to sustainability as an evolving process.

Lifelong learning and sustainability practice

For professionals in THE education is not confined to formal training. It continues in the contexts where sustainability is interpreted, applied and often contested. The complexity of sustainability challenges shaped by shifting environmental, cultural and economic conditions means that knowledge alone is not sufficient. Professionals are encouraged to develop the capacity to learn through experience, respond to ambiguity, and revise their understanding as new demands emerge.

Formal education remains important, especially when it provides foundations in ethical reasoning, systems awareness and contextual sensitivity. However, these capabilities mature over time. They take shape through engagement with real-world tensions and through reflective participation in professional settings. Lifelong learning in this context is not a supplement to practice; it is integral to maintaining sound judgement in unstable conditions.

This understanding is embedded in global frameworks that support sustainability education. UNESCO's Learning Compass defines learning as a dynamic, future-oriented process, one that supports navigation rather than the accumulation of fixed knowledge. Similarly, SDG 4, and particularly Target 4.7, frames education as a means of enabling responsible action through critical thinking, civic engagement and the development of relevant competencies.

Climate uncertainty, shifting policy landscapes, community priorities and supply chain reform all require professionals to remain agile in how they understand and apply sustainability concepts. In this regard, lifelong learning sustains credible, informed participation in sustainability work. It moves beyond aspiration by enabling new professionals to engage with uncertainty through continuous reflection and adjustment. For new professionals, this means staying alert to change, questioning familiar solutions, and remaining accountable to the contexts in which sustainability is pursued.

For those entering the THE sector, sustainability will not present itself as a single challenge or fixed direction. It will emerge through transition points that require adaptability and consistency in equal measure. Lifelong learning supports this kind of responsiveness. It enables professionals to refine decisions without losing direction, and to act constructively in conditions where certainty is not available.

Professional development is a continuing commitment that evolves alongside practice. In the context of sustainability, this commitment is transformative. It enables professionals to question assumptions, adapt approaches and remain responsive as their responsibilities evolve, contributing to personal and professional emancipation.

Wellbeing and the art of sustainability

> **INDUSTRY VOICE: STEVEN ENGLAND**
>
> Steven England, Director and Principal Consultant Facilitator at The Art of Sustainability, states 'The sustainability journey begins within. In my work at The Art of Sustainability, we see how mindfulness, creative inquiry, and self awareness are not soft skills, but core competencies for navigating complex change. Through the Wellness of Being® framework, we help individuals reconnect with their inner compass, supporting wellbeing not as an outcome, but as a foundation for ethical leadership. Sustainability isn't something external to act upon; it's a way of being that requires presence, courage, and reflection.
>
> 'Cultivating this inner clarity enables professionals to engage with the world more openly and compassionately, and that's where meaningful transformation starts.'

Throughout this book, sustainability has been approached as a professional practice, one that draws on knowledge, ethical reasoning and the capacity to act in complex conditions. These dimensions of knowing and doing are necessary, but they are sustained through ways of being. Professional credibility in sustainability is shaped not only by decision-making, but by how professionals relate to themselves, to others and to the conditions in which they work. This section considers how wellbeing supports the longer-term capacity to engage ethically and effectively with sustainability challenges.

Wellbeing is often treated as a secondary concern in professional development. Yet in sustainability practice, it is a condition of sustained engagement. The ability to make sound decisions, support meaningful relationships and remain responsive in demanding contexts depends in part on the professional's own capacity to stay grounded and attentive over time.

The Art of Sustainability (AoS) offers one example of how this understanding can be embedded in professional culture. Through its Wellness of Being® (WoB) framework, it positions wellbeing not as a retreat from responsibility, but as part of how professionals stay attentive, effective and ethically present in their work. This approach foregrounds the sustainability of the self, sustained through mental, physical, emotional and spiritual wellbeing, as essential to long-term professional contribution.

In an industry often shaped by short-term outcomes, extractive business models and uneven labour conditions, this is not an optional concern. It is a necessary condition for ethical engagement. Professionals who attend to their own wellbeing are better placed to make clear decisions, to sustain relationships and to contribute meaningfully over time.

The AoS model reflects broader aims in sustainability education. It connects with the professional competencies explored throughout this book, especially ethics, systems awareness and futures literacy, and aligns with SDG 3 (Good Health and Wellbeing) and SDG 4 (Quality Education). It also reframes leadership as a practice of responsiveness rather than control, grounded in attention, not certainty.

For emerging professionals, this perspective suggests that sustainability is not only about external systems or institutional change. It is also about the internal capacity to meet challenges with care, steadiness and reflective intent. When well-being is understood as a shared responsibility of the organization and the individual, both the person and the practice benefit as performance improves and professional identity is strengthened.

An invitation to ongoing practice

This conclusion is not the end of the conversation. If sustainability is to remain relevant, it must be treated as a continuing process of inquiry, adjustment and shared learning. This requires revisiting assumptions, staying open to different perspectives, and engaging critically with tools and frameworks.

The material offered in this book is not intended as a fixed guide. It serves as a point of entry for deeper reflection. What does sustainability mean in your context? Who defines it? What trade-offs are at stake, and for whom? These are not abstract questions, they shape the decisions that will influence the future of THE sectors.

No single model or roadmap will resolve the complexity of sustainability. But meaningful direction can emerge through practice that is responsive and grounded in professional responsibility. Sustainability is not a destination. It is an active process shaped by context and maintained through deliberate engagement.

We hope this book has supported you not by offering final answers, but by encouraging a way of thinking, being, and doing that remains responsive to the increasing urgency of the sustainability crisis. That work must begin now.

11 | Extension material

Real-world examples and professional interviews

Introduction: Purpose and use of extension material

This final chapter brings together a range of materials that, while not included in the main body of the book, are relevant to its overall aims. These include real-world case examples and professional interviews that build on themes covered in earlier chapters. In some cases, the material reflects valuable contributions from external collaborators or industry practitioners that did not align neatly with a particular chapter's focus. In others, the content provides a deeper exploration of specific issues that are better presented as standalone resources. Rather than being omitted, these materials are included here as a dedicated resource for students, lecturers and early-career professionals.

The extension material is organized into two sections:

- Extended real-world examples provide additional examples of sustainability practice from different cultural, organizational and geographical contexts. These include community-led waste management projects, industry-driven sustainability programmes, and initiatives addressing social equity in aviation and urban settings.

- Professional interviews offer reflections from individuals working in sustainability-related roles. These conversations focus on topics such as professional development, organizational change and the use of creative or systems-based approaches in sustainability work.

This chapter can be used in a number of ways. Lecturers may wish to incorporate materials into workshops, tutorials or assessments, while students may find them useful for independent study or group discussion. Each entry includes a brief note explaining its connection to the main text and suggestions for use. The inclusion of this chapter reflects the book's wider intention to support learning that is both conceptually informed and grounded in practice. These materials are intended to complement the main chapters and offer further entry points into sustainability in tourism, hospitality and events (THE).

Extended real-world examples

This section presents additional real-world examples that complement and extend themes explored in earlier chapters. Each example has been selected for its relevance to sustainability in THE, with a particular emphasis on lived experience, community-led solutions and sector innovation. Where applicable, the real-world examples include a brief reference to chapters where related concepts or issues were introduced.

REAL-WORLD EXAMPLE Bulon Island Community Waste Bank Project (Thailand)

Related themes: Community-based tourism, waste management, poverty and infrastructure (see Chapter 2).

This real-world example focuses on Bulon Island, a small fishing community in southern Thailand, and its response to increasing waste generated by tourism. The island lacked formal waste management infrastructure, and open burning and dumping were common practices. With support from local leaders and academic partners, the community initiated a Waste Bank project aimed at promoting environmental awareness and introducing economic incentives for recycling. Although the project faced logistical and sustainability challenges, it offers insights into how tourism can both stress and strengthen local environmental systems.

The Bulon Island community in Satun Province in southern Thailand represents a traditional fishing community. The inhabitants are mostly Muslim sea gypsies, an ethnic minority group also known as Chao Lay. Chao Lay began to permanently settle on Bulon Island a few decades ago. Sea gypsies on Bulon Island have minimal experience in community development because the geography itself is far from the mainland. The inhabitants resorted to traditional fishing to make a living. By using both monetary and non-monetary poverty measurements, it can be said that the Bulon community is living in poverty. Their living conditions are also poor due to the lack of electricity, clean drinking water or access to basic education. Local people on Bulon Island often cannot make ends meet as their incomes are only enough for daily expenses, and they normally end up being in debt to lenders or wholesale fish buyers.

As a result, managing waste was not a big concern, yet it has become one of the biggest challenges for the community. Bulon Island does not have a proper waste management system. The commonly used methods include open burning, open dumping or digging up the ground at the back of houses. The researcher's observations during the study revealed that every household has its own open burning area, some in the backyard and some in front of their house. When asked about it, one of my research participants stated, 'My priority is to feed my family, so I don't know, we just burn it and it's done'.

The concern with open burning and open dumping is air pollution and hygiene, because the community uses rainwater for consumption, which means the open dumping method may cause health and sanitary problems. The average waste disposal in Bulon Island is about 550 kg per day during the tourist season, and 150 kg per day during the monsoon season when the island is closed to tourists. The disposal includes household waste, which comprises food waste, glass and plastic bottles, fishing nets and other fishing equipment.

During the study, the researcher worked closely with the community, local government and tourism enterprises on the island to introduce the Waste Bank project. The researcher and the community spent two years putting the project into place. The first year was to raise concern about the disposal problem, especially in tourism areas, and encourage the local community to address this by first teaching them to categorize waste disposal and how to manage each type.

The local government cooperated with the Prince of Songkla University to create biogas from food waste from bungalows and households. The biogas project is located at the school and was started in 2012 with the aim of encouraging the local community to utilise food waste. However, this project is not operating currently because there is no one to look after it, and the locals do not see the benefit of this project. The lesson learned from this project was that the local people do not see the benefit, and it is not convenient for them.

The Community Waste Bank Project initiative came about as a result of the focus group run by the researcher which involved community leaders, local people, tourism enterprises and tourists. The community leader, Mr Cha, said: 'How about we buy their waste?' Later, locals shared their ideas of what they wanted in return.

The Community Waste Bank Project serves multiple objectives: it reduces the burden on municipal waste management systems, promotes environmental awareness, generates economic opportunities and enhances community cohesion. By turning waste into a valuable resource, the project aligns with the principles of the circular economy, emphasizing the importance of recycling, reusing and reducing waste. The rubbish can be sold or exchanged for something else, such as cooking oil, sugar or fish sauce, at the waste bank, which a community leader currently runs. The problem with this programme is that there is insufficient space for collecting the rubbish and locals make only a little profit because they have to collect the rubbish to sell on the mainland after transporting it in their own boats, which incurs high fuel costs.

In conclusion, poor social amenities (waste management) are part of the local people's perception of poverty. Tourism has brought indirect benefits through the improvement of waste management, increasing the attention of local governments to put more money into the system and improve important basic infrastructures; local people felt that the improvement of basic household services, such as water and waste management and the electric generator, have derived from the development of tourism in the area.

Source: Hunt, 2017

REAL-WORLD EXAMPLE   ReciclAção in Rio's favelas (Brazil)

Related themes: Environmental education, grassroots sustainability, community resilience (see Chapters 2 and 4).

Based in the favela of Morro dos Prazeres, the ReciclAção project was developed in response to a deadly landslide caused by unmanaged waste. Local young people, supported by NGOs, created a programme of environmental education, waste sorting and reuse. The project's emphasis on community ownership, local identity and informal leadership demonstrates how sustainability initiatives can emerge to generate meaningful social and environmental change under challenging conditions.

ReciclAção (RecyclAction) is an educational environmental project which attempts to maintain hygiene and cleanliness throughout the community to improve its environment. The project was initiated through Prazeres' community-based organization Prevenção Realizada com Organização e Amor – PROA (Prevention Realized Through Organization and Love). PROA focuses on health issues, which include important issues of sexual health, sensitizing the community to health through education (Pizzimenti, 2017).

ReciclAção also focuses on individual health through physical wellbeing. The project was a response to the community's insurmountable accumulation of waste. Sadly, in April 2010, the build-up of waste was a contributing factor to a devastating landslide which killed more than 30 people and destroyed the community structure (Cade, 2024). This instigated the coalition among community activist groups, young people from Prazeres and non-governmental organizations (NGOs), including UNICEF, monitoring and mapping the level of waste deposits in ten favelas (Pizzimenti, 2017). From this, 25 young people from Prazeres realized that the community needed to start an initiative that was focused on raising awareness about waste. The key focus was on how it was produced, the associated risks of waste, and how to dispose of it more consciously. However, it was extremely difficult for them to identify current organizations that focused on the underlying cause of waste – a lack of environmental awareness.

Cris dos Prazeres (a founder of ReciclAção) defines it as 'a social technology that orients human beings towards a different, more conscious way of seeing the environment, always from within [the favela] to the outside'. The project's mission is 'To orient human beings towards a different, more conscious way of seeing the environment through environmental education and recycling'.

Education is the key focus of this project to raise awareness of protecting the environmental surroundings as something that belongs to them. ReciclAção operates various meetings and discussions in Prazeres to proactively promote awareness, discuss challenges and generate new ideas. Their poster read, 'The trash you throw on the ground doesn't talk, but it says a lot about you!'

The project has a triangulated approach in dealing with waste through the three Rs of reducing, reusing and recycling. The team set up 'eco-points' across Prazeres with waste bags for the residents to throw their rubbish. These bags are sorted daily and separated into recyclable and non-recyclable materials for distribution to the correct location. Additionally, ReciclAção hold workshops on how to repurpose materials for transforming them into objects to sell or use, teaching people the economic value of this. By setting examples, the community has gradually started to encourage others to follow the movement, who naturally contribute and dispose of waste using the eco-bags, bringing social change (Pizzimenti, 2017).

The lack of access to basic resources seems to empower favelas to develop and implement their own projects. Through such projects, these culturally-rich communities can educate others on environmental issues, raising awareness for future sustainable change. The project leads direct educational initiatives through schools and universities, even in the United States. These educational ventures contest the government's lack of care for sanitation and environmental health in favelas. ReciclAção works to connect residents with the environmental spaces around them to encourage and educate them on the value of caring for collective spaces for a positive effect on their futures.

However, many issues such as ongoing violence and criminal group conflict and violence make it problematic for the clean-up operations to take place. During 2015, the increased violence and shootings resulted in the clean-up being postponed, but it did not prevent residents from disposing of their waste in the eco-bags, enhancing the value of the ReciclAção project.

It is evident that such projects may bring many benefits to one community, educating others, raising awareness of such imminent and important issues facing the world's climate contributing to and promoting sustainable development.

(Source: Cade, 2024)

REFLECTIVE QUESTIONS

1 What benefits can grassroots projects bring to disadvantaged communities?
2 What are the social benefits of these grassroots projects?
3 In what other ways could communities like Rio's favelas generate positive social change for their future?

REAL-WORLD EXAMPLE	Gender equality and EDI at Virgin Atlantic (UK)

Related themes: Social sustainability, gender, aesthetic labour, organizational change (see Chapter 4).

This real-world example analyzes Virgin Atlantic's attempts to address gender inequality and promote inclusivity through uniform policies, recruitment practices and branding. Drawing on feminist critiques of aesthetic labour and organizational norms, it explores how corporate policy can both challenge and reinforce gender expectations. It also considers the limits of such initiatives in the context of international operations and cultural constraints, such as the suspension of gender-neutral uniforms during the Qatar World Cup.

Key concepts:

- gender equality
- aesthetic expectations
- policy development
- organizational initiatives

The THE industry serve as a powerful mechanism for advancing social sustainability, a concept defined by the World Bank (2020) as the imperative to 'put people first' in development processes. This approach seeks to promote the social inclusion of marginalized and vulnerable groups by empowering individuals, fostering cohesive and resilient communities, and ensuring organizations are both accessible and accountable to all stakeholders. Social sustainability aims to meet socio-cultural needs fairly and equitably, promoting equality and social justice as essential steps toward building more inclusive and equitable societies.

Aviation (i.e. air travel) plays a crucial role in tourism, as it facilitates global connectivity and contributes to economic growth (Papatheodorou, 2021). While a great deal of progress has been made toward gender equality in aviation, development remains gradual and imbalanced, with progress varying across different areas of the sector. This real-world example focuses on Virgin Atlantic, a British full-service airline with a distinctive brand identity and a willingness to challenge industry norms. The aim of this example is to present an overview of the gender dynamics in the aviation industry before offering a critical review of Virgin Atlantic's attempts to incorporate EDI principles and practices within the organization.

Gender is a critical framework through which various social inequalities and exclusions are manifested and recognized (Clarke and Braun, 2008). Social inequalities are shaped by 'gender essentialism,' the belief that men and women possess fundamental, unchangeable qualities that inherently differentiate them. For instance, while men reflect aggressive traits and behaviours of dominance, women are considered emotional and passive (caregiving characteristics) (Matthaei and Brandt, 2001). The assumed view of women as nurturers is deeply embedded in the common belief that women are more suited than men to provide service and care (Eckert and McConnell-Ginet, 2013).

Airline organizations play a vital role in maintaining rigid gender norms. Men in aviation predominantly occupy authoritative positions of power and skills (e.g. captain, pilot, engineer). Meanwhile, women primarily occupy roles that reflect home-care responsibilities (cabin crew, ground crew and customer service). Hochschild's (1983) and Simpson's (2014) work has shown that the cabin space is regarded as a highly feminine space where women 'doing' service is viewed as an extension of domesticity and femininity, traits seen as a natural and essential part of 'women'.

Virgin Atlantic, founded in 1984, has pioneered a range of innovative and distinctively novel organizational structures and approaches, setting new service standards. Nevertheless, like many other airline companies, Virgin Atlantic has emphasized the importance of women's physical appearance and sexuality in representing its brand. Previous representations are noted in its 25th anniversary 'tongue-in-cheek' advertising campaign 'Still Red Hot' which depicts sexist and derogatory attitudes towards women (Duffy, Hancock & Tyler, 2017) – producing symbols or messages that promote narratives that maintain harmful stereotypes, further perpetuating the role of the female flight attendant, limiting individuality and self-expression. The representation of the female flight attendant has historically been of the conventionally attractive, slim and glamorous 'woman'. Airlines emphasized aesthetics, favouring a pleasing exterior over cognitive ability in cabin crew work (Bor and Hubbard, 2006; Gertel, 2014). The presentation of cabin crew, particularly female flight attendants, is often used to present the desirable characteristics and 'beauty' standards.

As attitudes towards sustainability progress, airline organizations are increasingly moving towards embedding EDI principles by implementing initiatives and policies that serve as guidance and organizational frameworks. In 2019, Virgin Atlantic modernized its gender identity policy by changing uniform codes and make-up regulations for both men and women: 'Not only do the new guidelines provide more comfort, they empower our team with more choice on how they want to express themselves.' (Virgin Atlantic, 2023)

The changes implemented by Virgin Atlantic illustrate how industry-specific policies can align with evolving cultural understandings of gender, promoting more significant equity and setting the stage for future advancements. Through such initiatives, the aviation industry is a case study of how organizational settings can lead the charge for meaningful social change. Since 2019, Virgin Atlantic has continued to challenge aesthetic expectations governing the management of airline cabin crew (The Conversation, 2019; Virgin Atlantic, 2022). Its 2022 'Be Yourself' campaign reflects its workforce and customers' diversity. The airline's bold industry-leading inclusivity initiatives contribute to reshaping certain naturalized assumptions of gender that exist within the aviation sector.

(Source: Vernes, 2023)

> **REFLECTIVE QUESTIONS**
>
> 1 How do Virgin Atlantic's 2019 updated uniform and make-up regulations contribute to addressing broader inequalities beyond the aviation sector? Discuss whether issues of EDI are present in related sectors such as THE.
>
> 2 What roles do airline organizations play in offering its employees more choice and flexibility in self-expression and individuality?
>
> 3 Propose strategies the airline can implement to make their EDI policies and initiatives more resilient to external challenges in the future. How can airlines maintain momentum and advance its EDI initiatives?

Professional interviews

This section features interviews and reflective conversations with professionals working in sustainability-related roles across different sectors. These contributions are intended to offer students and early-career practitioners insight into real-world challenges and strategies that may not be visible in academic or policy literature. Each perspective also illustrates how sustainability knowledge and skills are applied in practice and can help inform career planning.

Interview 1: Miranda Simmons – Sustainable production in the film and TV industry

Interviewed by Michel Mason.

Related themes: Systems thinking, logistics, cross-sector learning (see Chapter 3 and Chapter 5).

Miranda Simmons is a sustainable production advisor, consultant and trainer working across the UK's film and television sector. In this interview, she reflects on her work with BAFTA Albert and other industry organizations focused on reducing the environmental footprint of media production. Key themes include energy use, transportation, local sourcing, staff training and cultural change. Students are invited to consider what the THE sector can learn from these parallel sustainability efforts, particularly in terms of coordination across short-term, project-based teams.

IN CONVERSATION WITH MIRANDA SIMMONS

In this insightful interview, Miranda Simmons shares her views and insights on driving environmental sustainability in the media, tackling industry challenges, and paving the way for a greener future in the film and television industry.

Michel: Miranda, thank you for joining us! To start, could you introduce yourself and tell us about your work?

Miranda: Of course! I'm Miranda Simmons, a freelance sustainable production advisor, consultant and trainer for media organizations.

One of my training clients is BAFTA Albert, the sustainability organization for the film and TV industry. I also work with a TV production company specializing in unscripted international factual programming and a sustainable production consultancy called Picture Zero. Additionally, I've created training presentations for Media Cymru and All Spring Media, which offer courses for the industry.

Michel: BAFTA Albert sounds intriguing. Could you tell us more about it?

Miranda: Absolutely. BAFTA Albert, or just Albert, is the leading organization in the screen industry for environmental sustainability. They've been working with film and TV companies to reduce their environmental footprint and inspire a sustainable future through the content they create.

Albert is owned by BAFTA and backed by the industry, which gives them a robust platform to drive meaningful change. Their mission is to empower screen professionals to identify and act on opportunities for sustainability, whether on-screen or off-screen.

Michel: What are some key sustainability initiatives that Albert is currently focusing on?

Miranda: Albert has a variety of initiatives. They've introduced the Studio Sustainability Standard and the Climate Content Pledge. They've also partnered with ScreenSkills to provide online department-specific training for crew and offer sustainable production workshops that guide production teams in calculating carbon footprints accurately.

A key area of focus is encouraging productions to hire local crews, which can reduce the environmental impact of travel and support regional economies. Albert is also promoting professional development for staff, helping build a knowledgeable workforce capable of implementing sustainability measures effectively.

Supplier engagement is another focus area, including working with catering providers to encourage greener practices. Across the sector, a key priority has been embedding specialist sustainability staff within productions, as sustainability responsibilities are often assigned to junior team members who may lack the expertise to make a real impact.

Michel: Can you share specific projects that showcase Albert's commitment to energy efficiency and renewable energy?

Miranda: Certainly. One standout initiative is their Creative Energy scheme, which partners with a 100 per cent renewable energy supplier. Even BAFTA itself is now on a renewable energy tariff.

Beyond that, the industry is exploring hybrid battery power for off-grid locations to reduce reliance on diesel generators. Trials with energy tracking software are also underway, helping productions feel confident about switching to smaller generators or battery alternatives.

Michel: What are some of the challenges and impacts you've seen with these initiatives?

Miranda: Carbon footprints in the industry have rebounded since the lockdowns, but they're not as bad as 2018 levels. A significant impact has been the improved accuracy of carbon footprint data and the inclusion of sustainability clauses in UK broadcaster commissioning agreements. This means productions must consider carbon footprints and action plans from the outset.

However, challenges remain. Normalizing sustainability in a sector largely made up of freelancers and small businesses, where every production starts from scratch, makes it tough to establish consistent practices.

Culture change is difficult. Things have been done in the same way for decades. There are concerns over who's paying for green technology or the extra time needed – and who will be responsible if something goes wrong. Economic pressures haven't helped either, with many people out of work and reluctant to adopt green workflows when on a short-term contract.

Other barriers include resistance from lower-paid suppliers, the logistical issues of transportation, and the influence of American funders, who lag behind Europe in sustainability priorities.

Michel: Transportation is a significant issue in production. How does Albert address this?

Miranda: It's a tough one! Travel is the biggest carbon emissions impact – and the hardest one to crack. Production often involves long and erratic schedules and operate in isolated locations making some sustainable options such as electric vehicles problematic.

Albert is involved in trials of crew travel surveys to better understand their needs and openness to alternative travel options. Car sharing is being encouraged, but it's not always practical given varied schedules.

Hiring local crews is a significant step forward here. Not only does it cut down on transportation emissions, it also helps develop the skillsets of professionals in different regions.

For younger crew members who don't drive, transportation is becoming a bigger challenge. There are also trials with portable EV chargers on location, which could ease concerns about electric vehicles.

Michel: What about circularity? Are there initiatives for recycling sets, costumes and props?

Miranda: Recycling is often the go-to solution right now, though some businesses like Prop-Up Project connect productions with schools or charities to repurpose props. Community Wood Recycling is another great example – they refurbish and collect old set pieces.

On long-running series, some productions use asset managers to store and resell items. However, storage costs and concerns about intellectual property can be barriers.

In costume design, it's tricky because productions often require multiple identical outfits, and designers on high-end shows might prioritize creative awards over sustainability. That said, professional associations are pushing for more circular designs, and the mantra 'Rent and Return, not Buy and Bin' is gaining traction.

Michel: What's next for Albert and the wider industry in terms of energy efficiency and renewable energy?

Miranda: The big push is to electrify everything, from battery-powered sets to securing mains connections at studios and locations. Energy tracking and battery trials are essential for reassuring crews that these changes can also bring cost savings.

International adoption of Albert's standards would be a game-changer, creating consistency across the board. Increased training, engagement and collaboration with funders to support green production budgets are also on the horizon.

Michel: What advice would you give to production companies looking to improve their sustainability?

Miranda: Start by hiring local crews wherever possible – it's one of the easiest ways to reduce emissions and strengthen regional talent pools.

Fund trials and gather data to prove greener options work. Build a timeline for full electrification on set by a specified date and make sustainability part of early planning conversations.

Senior decision-makers need to take ownership, ensuring greener practices aren't left to junior staff. It's about making sustainability a shared responsibility across the production team.

Interview 2: Steven England – The Wellness of Being® and sustainability education

Interviewed by Michel Mason.

Related themes: Professional development, creativity, emotional learning (see Chapters 3 and 6).

Steven England is director and principal consultant facilitator at The Art of Sustainability® and the founder of The Wellness of Being®, a methodology that blends systems thinking, emotional intelligence and creative facilitation to support sustainability learning. In this conversation, Steven discusses how the arts, nature connection and embodied practices can enhance professional development and personal sustainability. His reflections offer an alternative view of sustainability education, one that moves beyond compliance or metrics and toward personal transformation.

This interview may be of particular interest to educators and facilitators working in student wellbeing, leadership development or cross-disciplinary contexts.

IN CONVERSATION WITH STEVEN ENGLAND

In THE, sustainability is often framed in terms of environmental policies, carbon reduction and responsible business practices. However, sustainability begins with the self. In this interview, Steven England argues that professional development for sustainability is not just a set of technical solutions but a deeply personal and professional practice that requires mindfulness, emotional intelligence and systemic thinking.

Michel: Steven, thanks for joining me today. Sustainability in THE is often seen as a technical or policy-driven issue. But you argue that it begins with the Self. Can you explain what you mean by that?

Steven: Absolutely. Sustainability isn't just about environmental actions or policy frameworks – it's about how we, as individuals, show up and engage with the world. That's why at The Art of Sustainability – AoS – we emphasize that 'Sustainability begins with your Self'. And this is something critical – there's a big difference between yourself and your Self as two separate words.

Most people think of 'yourself' as just who they are in their daily life, their job, their responsibilities, their routines (as the selfish ego). But when we talk about Self with a capital 'S', we are talking about something deeper – the conscious, the present and connected part of you.

Michel: And why is this distinction important from a professional point of view?

Steven: This distinction is vital for sustainability professionals because sustainability is not just about external actions – it's about internal awareness. If we only approach sustainability as something 'out there' to fix, we miss the fact that our mindset, our emotional resilience and our presence in the work matter just as much as our knowledge.

That's why we emphasize that 'Sustainability begins with your Self.' If we want a sustainable THE industry, professionals need to develop a good practice mindset and behaviours alongside their professional skills. This means understanding that sustainability is a living, evolving practice – not just a box-ticking exercise. It's about being present, resilient, and creative in how we approach challenges.

Michel: That's a very different take on professional development. How does this translate into practice for industry trainees?

Steven: It starts with recognizing that no one person or organization has the full picture of sustainability. It's like a jigsaw puzzle, where each of us holds a unique piece. The challenge is how we bring these pieces together – and that begins with self-awareness and having emotional intelligence.

That's where The Wellness of Being (WoB) methodology comes in. It's an emergent methodology – sustainability is the framework where the WoB methodology can be applied. It emerged out of participant collaboration in The Wellness of Being Programme which I formalized and trademarked. It's designed to help professionals build self-awareness and resilience. Instead of looking at sustainability as a theory, we look at how people practically engage with it – what they feel, how they act and what influences their choices.

We also draw on arts-based research that uses creative methods – like visual art, storytelling and performance – to explore these ideas. It helps us make sustainability more personal and engaging. This is crucial in professional development because learning isn't just about facts – it's about how we internalize and act on knowledge.

Michel: And how is that achieved through professional development?

Steven: We use a combination of meditation, mindfulness and reconnecting with nature. It's not about compartmentalizing wellbeing as a separate intervention into a two-hour workshop. It's about integrating these practices into professional development; we're not just teaching sustainability as a concept instead it's about embedding [embodying] wellbeing into professional growth. It's an approach that shapes professional competencies.

Michel: Can you expand upon that a bit more?

Steven: Yes, what we are talking about are skills and competencies in self-awareness and emotional intelligence; in mindfulness and patience in decision-making; resilience in handling sustainability challenges; and the ability to think both systemically and creatively. It's only when we develop these skills can we, as professionals, navigate the polycrisis with agency.

Michel: Can you explain what you mean by polycrisis?

Steven: Of course. Polycrisis refers to multiple interconnected crises – climate change, biodiversity loss, inequalities – all happening at once, amplifying each other. The UN Sustainable Development Goals – the SDGs – illustrate this well; they show that issues like poverty, health and environmental sustainability are deeply linked. Solving one in isolation won't work – we need systemic, holistic approaches that address the bigger picture.

Michel: One aspect of your methodology we have not touched on is mindset. How do you define a growth mindset and how does this fit into Wellness of Being?

Steven: First, we have to ask ourselves, 'What is it we are trying to make well?'. The answer is our Being – our Self – the quality of our experience.

While all professional development aims to bring about change, with an experiential growth mindset, it is the trainee themselves who drives that change through experience, reflection, and adaptation, rather than just absorbing knowledge.

A good example is our work with the Carbon Literacy Project, which provides professionals with practical knowledge on climate change and carbon footprints. Rather than deliver it as a lecture or scientific download, we use six key mindfulness techniques: permissioning, concentration, returning, patience, enjoying, and letting go to help participants experience sustainability rather than just learn about it.

If we learn how to reflect upon our experiences, we learn how to adapt. We can also learn how to see sustainability as an opportunity to innovate rather than a crisis to mitigate. What if sustainability training shifted from problem-solving to potential-realizing? That's the shift professionals need if they are going to bring about organizational behavioural change.

Michel: Yes, but in a corporate or commercial setting, there's always the question of measurement. How do you demonstrate the impact of these approaches?

Steven: You're right, and it's something we've thought about. Wellbeing is about the quality of experience, but we recognize that organizations need measurable outcomes to justify investment. One way is through social prescribing, which is a healthcare approach that connects people to non-medical interventions like the arts, nature or mindfulness. One example is the World Wellbeing Movement Measurement (WWM) which is a commercial tool linking wellbeing to business performance. Tools like the WWM help bridge the gap between experiential approaches and hard data.

Michel: You've given us a lot to think about. If you had to leave future THE professionals with one key takeaway, what would it be?

Steven: I'd say: Sustainability begins with your Self. If you want to make an impact in this industry, start by looking at how you show up, how you engage with challenges and how you take care of your own wellbeing. The more present, creative and resilient you are, the more you can contribute to a truly sustainable tourism industry.

EMBODYING SELF – REFLECTIVE QUESTIONS

Steven England talked about embodying Self and reflection as a way to grow as a person and professional. Below are some effective reflection techniques that you could apply to your personal and professional development.

After each reflection, ask: How did that make me feel? How can I extend this feeling of wellbeing to the whole of life?

1 Setting aside a few moments each day to sit quietly and focus on your breath can help calm your mind and bring clarity of thought. Consider participating in mindfulness via an online or in-person session.

2 A mindful walk offers a chance to connect with both nature and your inner thoughts. As you walk in nature, pay attention to your surroundings, the sensations in your body and the thoughts that arise.

3 Imagine your own wellbeing and how interconnected and interdependent it is with friends and loved ones. Have open conversations to share your aspirations and challenges while inviting constructive feedback. These discussions can help you recognize strengths and areas for improvement you may not have noticed.

Concluding reflection

This final section brings together a diverse range of applied insights that build on the conceptual and practical foundations established throughout the book. It showcases real-world examples of sustainability in action, from community-led waste initiatives to large-scale food service programmes, and from professional development through mindfulness to inclusive reforms across industries.

Readers are encouraged to revisit this chapter for inspiration, practical examples or alternative perspectives that support learning and strategic planning within their own contexts. While sustainability challenges are complex, they also present opportunities for growth through reflection, collaboration and innovation.

To further support engagement with the book's themes, a suite of online resources is available in the form of student activities. These extension materials offer additional opportunities to apply key concepts, including tasks that encourage learners to explore how sustainability strategies can be adapted to roles within THE.

Further resources

McCarthy, F, Budd, L and Ison, S (2015) Gender on the flightdeck: Experiences of women commercial airline pilots in the UK, *Journal of Air Transport Management*, 47, pp 32–38

Smith, W E, Cohen, S, Kimbu, A N and de Jong, A (2021) Reshaping gender in airline employment, *Annals of Tourism Research*, 89 (103221)

Thackray, L (2022) Virgin Atlantic isn't offering new gender-free, inclusive crew uniform on World Cup flight, The Independent, 15 November, www.independent.co.uk/travel/news-and-advice/virgin-atlantic-inclusive-uniform-world-cup-flight-b2225595.html (archived at https://perma.cc/K5NA-X8FN)

References

Bor, R and Hubbard, T (2006) *Aviation Mental Health: Psychological implications for air transportation*, Routledge, Oxford

Cade, N (2024) Mega-event fragmentation of Rio's favelas: The socio-cultural trauma of Morro dos Prazeres, PhD thesis, King's College London

Clarke, V and Braun, V (2008) Gender. In: D Fox, I Prilleltensky and S Austin (eds.) *Critical Psychology: An introduction*, Sage, London

Duffy, K, Hancock, P and Tyler, M (2017) Still red hot? Post-feminism and gender subjectivity in the airline industry, *Gender, Work and Organization*, 24 (3), pp 260–73

Eckert, P and McConnell-Ginet, S (2013) *Language and Gender*, Cambridge University Press, Cambridge

Gertel, A R (2014) Not just a pretty face: The evolution of the flight attendant, senior Honors project

Hochschild, A (1983) *The Managed Heart: Commercialisation of human feelings*, University of California Press

Hunt, B (2017) Poverty reduction and tourism in Bulon Island Thailand, PhD thesis of Dr Banthita Hunt

Matthaei, J and Brandt, B (2001) From hierarchical dualism to integrative liberation: Thoughts on a possible non-racist non-classist feminist future, Wellesley College Working Paper

Papatheodorou, A (2021) A review of research into air transport and tourism: Launching the Annals of Tourism Research curated collection in air transport and tourism, *Annals of Tourism Research*, **87**

Pizzimenti, S (2017) ReciclAção – RecyclAction – in Prazeres #SustainableFavelaNetwork [PROFILE], Rio On Watch, 17 October, https://rioonwatch.org/?p=39154 (archived at https://perma.cc/W4MY-A2UK)

Simpson, R (2014) Gender, space and identity: Male cabin crew and service work, *Gender in Management*, **29** (5), pp 291–300

The Conversation (2019) Why Virgin Atlantic's new makeup policy is mostly concealer and gloss, 8 March, https://theconversation.com/why-virgin-atlantics-new-makeup-policy-is-mostly-concealer-and-gloss-113211 (archived at https://perma.cc/LQW5-M9US)

Virgin Atlantic (2022) Virgin Atlantic updates gender identity policy, 28 September, https://corporate.virginatlantic.com/gb/en/media/press-releases/virgin-atlantic-updates-gender-identity-policy.html (archived at https://perma.cc/6KT6-EH72)

Virgin Atlantic (2023) Press Kit 2023, https://corporate.virginatlantic.com/content/dam/corporate/media-centre/Press%20Kit%202023.pdf (archived at https://perma.cc/ET7V-Z9TR)

World Bank (.n.d.) Social development, www.worldbank.org/en/topic/socialsustainability (archived at https://perma.cc/H77M-MFWS)

INDEX

à la carte menus 218, 247
Action Against Hunger 221
actionable insights 150
Advertising Standards Authority (ASA) 262, 265, 266, 267, 273, 274
agency 68
Agenda 21 2, 24, 39
agriculture 171, 203, 204, 205–06
Airbnb 105
airline sector 176, 177, 271, 273–74, 276–77, 304–05
Albert 61, 307–09
ambassadors 15, 223
Anderson's five Ps of professional development 69, 70, 71–78
animal experiences 275–76
anticipation-action-reflection cycle 69–70
anticipatory (future) thinking 20, 48, 63, 293
apps 183
Arrábida Natural Park 175–80
Art of Sustainability 296, 297, 310–13
ASA 262, 265, 266, 267, 273, 274
astrotourism 33
Athenaeum Hotel 160–61
attitudes 68
audit framework 149–50, 162–63
audits 146, 147–51, 159, 162–63, 165, 267
authenticity 101–03, 276
automated meter readings 152
avoidable food waste 201
'avoided' GHG emissions 147

B Corp 148, 157, 158
BAFTA Albert 61, 307–09
Bali 106
Barcelona 104, 105
barley production 207
beef-steak production 207
behavioural change programmes 152, 155–56
behavioural competencies 64–65
    see also collaboration; integration; problem-solving
biodiversity protection (loss) 82, 106, 170, 206
blue water footprint 205
bottom-up professional development 80–81
Brazil 142, 202, 302–03
Brighter Planet 273
Broadcast Committee of Advertising Practice 266
Brundtland Report (*Our Common Future*) 2, 23, 39
buffets 239, 247

'Build Back Better' framework 33
buildings infrastructure 153, 154–55, 176
    see also heating, ventilation and air conditioning (HVAC) systems
Bulon Island 300–02

carbohydrate food waste 211
carbon consciousness (green) claims 266–77, 280, 282–83
carbon credits 277
carbon dioxide (CO_2) emissions 15, 171, 183, 186, 202, 273–74, 276
carbon footprint measurement 144–47, 148, 171, 172, 201–05, 241, 268, 271, 308–09
carbon footprint reduction 151–56
carbon labelling 154
Carbon Literacy Project 312
carbon neutrality 276–77
carbon offset 277, 281
carbon sinks 172
cereal production 203, 206
certifications 115, 135, 148, 156–61, 165, 185, 187, 236, 239, 269
change programmes 151, 152, 155–56
chanoyu 94
Chao Lay 300–02
China 202, 205
circular economy 3–4, 169–73, 174–96, 204, 289, 309
Circular Economy (Scotland) Act (2024) 182, 183
Clearcast 273
client food requirements 220
climate change 60, 142, 170–71
closed loop system 174, 179
co-agency 68
    see also collaboration
Code of Broadcast Advertising 267
Code of Non-Broadcast Advertising and Direct & Promotional Marketing 267
coffee industry 142
cognitive competencies 63, 67, 68
    see also anticipatory (future) thinking; critical thinking; problem-solving; systems thinking
collaboration 20, 48, 64, 95, 106, 154, 188–94, 216, 280, 289
collective industry action 188–89
Committee of Advertising Practice 265, 266
community engagement 82
    circular economy 178, 180, 191–92

Index

cultural sustainability 95, 101, 102, 105, 110, 111, 116
   metrics 136, 137, 139
   see also local food
Community Homestay Network 191–92
Community Waste Bank Project 301–02
Community Wood Recycling 309
competency frameworks 44–46, 47–48, 56, 62–70, 292–93
Competition Act 270, *271*
Competition and Markets Authority 262, 263, 265–68, 270, *271*
Competition Bureau 270, *271*
competitor analysis 14
complexity, navigating 68
'compostable' claims *268*
Consumer International 263–64
Consumer Protection from Unfair Trading Regulations (2008) 267
consumption 205
   see also SDG 12
continuous improvement 70, 134, 156, 189, 247, 248
continuous professional development (CPD) 59–60, 70–71
Convention for the Safeguarding of the Intangible Cultural Heritage 27–28, 39, 90, 91–92, 97
Convention on the Protection and Promotion of the Diversity of Cultural Expressions 27–28, 39, 90, 92, 97
cook-to-order systems 217
core foundations 67
   see also cognitive competencies
corporate social responsibility (CSR) 111, 112–17, 123, 125, 132, *133*, 137–40, 149, 164, 186
Corporate Sustainability Reporting Directive 135, 270
CO$_2$-equivalence 202
CO$_2$ (carbon dioxide) emissions 15, 171, 183, 186, 202, 273–74, 276
Courtauld Commitment (Food and Drink Pact) 146, 213, 214
Covid-19 pandemic 209–10
Creative Energy scheme 308
critical thinking 20, *48*, 56, 63
cross-industry collaboration 188–94
cultural appropriation 99–100
cultural diversity 93
cultural heritage 91–96, 124, 125, 290–91
cultural native experiences 276
cultural sustainability 3, 21, 26–28, 30–31, 33, 82, 88–126, 276, 290–91
   see also Convention for the Safeguarding of the Intangible Cultural Heritage
culture, organizational 156, 290

Dark Sky Sanctuary 33
data analysis 14, *150*, 162–63
data gathering (collection) *150*, 162
decontextualization 100–01
Department of Conservation 114
deposit-return schemes 178
Destination Management Plan (Kaikōura) 33
developed (industrialized) nations, food waste 205, 207
developing nations 204–05, 207
digital brochures 179
digital menus 234
dining experience 247–48
direct (Scope 1) GHG emissions 144–45, 151–52, 156
disciplinary knowledge 67
'doggy' bags 223
driver training 155
Dubrovnik 104, 106
durability 185, 186

Earth Summit (UNCED) 24, 39
EarthCheck 32, 148, 157, 158, 160–61, 282
EAT-Lancet Commission 245
Eco-design for Sustainable Products Regulation 270
economic sustainability 1–2, 3, 31–32, 33, 36, 120, 135, 138, 237
ecotourism 176, 192, 275
Edge Hotel School 13–15
education 45, 110
   see also SDG 4
8Rs framework 175–81, 289
electrification 153–54, 155, 309
emotional foundations 67, 68
   see also self-awareness
employee attrition 70, 75
employee engagement 216, 280
employee satisfaction 136
EN ISO 14024 265
end-of-life management 187–88
energy consumption 145, 180, 184–85, 186, *190*
energy-efficient equipment 153, 249
Energy Star 185
environmental impact assessments 149, 176
environmental sustainability 2, 3–4, 29–30, 32, 235–37
   and cultural heritage 105–06, 109–10, 115–16
   metrics 134, 138
   professional development 68, 82
   see also Agenda 21; biodiversity protection (loss); blue water footprint; climate change; greenhouse gases (GHGs); land use; United Nations Environment Programme (UNEP)
epistemic knowledge 67
equipment disposal 146

equipment maintenance 249
equipment timers 153
equity principle 35
ESG 132–34, 140–43, 164, 272
   see also environmental sustainability; governance; social sustainability
ESG Framework for Tourism Business (UN Tourism) 135
ESG implementation 142–43
ethical business practice 68, 117–19, 125, 136–37, 138, 164, 185–86, 187
ethnocentrism 101
Eurocontrol 273
European Commission 40, 47–48, 94–95, 263, 264–65
European Cultural Tourism Network 94–95
European Supervisory Authorities 262
European Sustainability Competence Framework 40
European Union (EU) 40, 45–46, 204, 270, 271
   see also Green Claims Directive; Task Force on Climate-related Financial Disclosures
events 1, 110, 176–77
Extended Producer Responsibility 182, 187

F-gases 145
fair trade 239–40, 247
fairness 137
Fairtrade Foundation 185, 187
fake culture 101–03
FareShare 221
Federal Trade Commission 261, 270, 271
feedback loops 189
Feeding Britain 221
Financial Stability Board 135
'first-in, first-out' inventory management 250
five Ps (professional development) 69, 70, 71–78
five Ps (SDGs) 36
flamenco 94
flexible mindset 292–93
FLOCERT 185–86
follow ups (audits) 150, 163
food 201
food and beverage outlets 211, 239
   see also coffee industry
Food and Drink Pact (Courtauld Commitment) 146, 213, 214
food byproducts 204, 205
food loss 201, 204–05
food miles 238, 241–44
food procurement 154
food production (preparation) 206–08, 212–13, 215, 218, 221, 232, 241, 247–48
   operational processes 250–51
   portion control 220, 223
   sourcing 234–40
food wastage 201
food waste 146, 177, 183, 199–226, 236, 237

food waste costs 211
food waste disposal 221, 301–02
Food Waste Reduction Map 213–14
food waste reduction strategies 179, 212–23, 242–44, 249, 277
forecasting tools 217
Forest Stewardship Council (FSC) 185
fossil fuels 171, 177, 185
freezer efficiency 250
future (anticipatory) thinking 20, 48, 63, 293
Future of Education and Skills 2030 66

GDP 3
gender equality 304–05
Ghana 94
Glasgow Declaration on Climate Action in Tourism (2021) 171
glass recycling 179, 211–12, 219, 243
Glastonbury Festival 190, 191
Global Code of Ethics for Tourism 136
Global Compact 140
Global Sustainable Tourism Council 279
global warming potential options 145
goal (objective) setting 143, 251
Gold Standard 281
governance 105, 106–11, 138, 148, 149, 294
   see also Corporate Sustainability Reporting Directive; Sustainable Development Goals (SDGs)
Green Breakfast Initiative 147–48
green building certification 154
green claims 266–77, 280, 282–83
Green Claims Code 270, 271
Green Claims Directive 264–65, 270, 271, 282
Green Fins 279
Green Guides 261, 270, 271
Green Key 148, 157, 158
Green Pearls® 81–82
GreenComp 40, 47–48
Greenhouse Gas Protocol 144
greenhouse gases (GHGs) 3, 144–47, 170–72, 182–85, 191, 202–03, 205, 206, 209, 241
   see also CO_2 (carbon dioxide) emissions
greenhushing 281
Greenpeace 262
greenwashing 42, 260–84
Greenwashing Academy Awards 262
greenwashing prevention strategies 279–81
grounded sustainability 61–62
growth mindset 58, 312
growth strategy 136
$GtCO_2eq$ 202–03
Guardians of Grub 146, 214, 222
Guidance on Environmental Claims on Goods and Services (CMA) 265–66

Hague Declaration 23–24, 39

Halstead, Rowen 242–44, 246
Hawaii 100
health foundations 67
    see also wellbeing
heating, ventilation and air conditioning (HVAC) systems 147, 152, 153
heritage 92
Hilton Hotels 139, 147–48
homestays 191–92
hospitality 1, 70, 138, 146, 209, 277–278
    see also Green Key; hotel sector
hot water management 153, 176, 177
Hotel Casa Palmela 175–81
hotel sector 135, 175–81, 277–78
    see also Edge Hotel School
household food waste 204, 208–09, 210

IceHotel 115–17
IFRS Foundation 135
imperfect produce 237, 243
inclusivity 35, 36, 95, 122, 304–05
indirect GHG emissions 144, 145–46, 152, 185
    see also Scope 3 GHG emissions
individual training needs analysis 79, 84
induction training 78
industrial food waste 204
Industrial Revolution 173
industrialized (developed) nations, food waste 205, 207
intangible cultural heritage 91–92, 93, 94, 95
integration 35, 48, 64
interdisciplinary knowledge 67
intergenerational cultural knowledge 102
Intergovernmental Panel on Climate Change (IPCC) 171
International Consumer Protection and Enforcement Network 263–64
International Union of Official Travel Organizations (IUOTO) see UN Tourism (UNWTO)
International Whale Commission 114
inventory management 250, 277
ISO 14001 148, 157
ISS 223

Japan 94, 202
job training 78
just-in-time delivery 218

Kaikōura earthquake 31–33
Kaikōura Social Recovery Plan 32
Keep America Beautiful 261
Kente weaving 94
Key Competencies for Sustainability (UNESCO) 47–48, 63–66
kitchen gardens 238
knowledge 67, 93

labour practices 1, 3, 30, 82, 134, 138, 149, 186, 187
land use 205
Learning Compass 66–70, 295
leasing models 174
LED lighting 152, 176
LEED 148, 154, 157–58
legislation 136, 148, 159–60, 271–72
    see also Circular Economy (Scotland) Act (2024); Competition Act; Extended Producer Responsibility
life cycle assessments 146, 154, 185
lifelong learning 59–60, 295–96
linear economy 172–74
local crews (staff) 309
local excursions 179
local food 189, 235–40, 241, 247, 251–52
'low-hanging fruit' (quick wins) 151, 152–53

maintenance 187
make-to-order systems 217
management plans 82
Manchester Hoteliers Association 209
Māori protocols 28, 33, 276
marae 276
Marriott Hotels 139
mass tourism 103, 109
McKinsey & Company 106–11
McMaster, Douglas 242–44
meat production 203, 207
Mediterranean diet 94, 95
menu layout (design) 240–41, 247
menu planners 233, 247
menu planning 217–18, 219, 222, 230–55
menu presentation 247–48
    see also portion control
metacognitive skills 68
methane 182–83, 202, 203
metrics (measurement) 22, 40, 49, 131–65, 190, 215–16
    carbon footprint 171, 172, 201–05, 241, 268, 271, 308–09
    see also Statistical Framework for Measuring the Sustainability of Tourism
mussels 239

national parks 178–79
national skills partnerships 45
nationalism 101
natural heritage 92–93
natural resources 205–08
    see also fossil fuels
Nepal Tourism Board 192
net positive 191
net zero 171, 172
Net Zero Carbon Events Roadmap 151
New Zealand 28, 31–33

Next Tourism Generation Alliance 40, 45–46
Ngōti Kuri iwi 32
normative competency 48, 56, 64
North Canterbury Transport Infrastructure Recovery alliance 33
nudges 155

objective (goal) setting 143, 251
OECD Learning Compass 66–70, 295
operational training needs analysis 79, 84
organic fibres 186, 187
organic produce 236, 237, 238, 239–40, 247
organizational culture 156, 290
organizational training needs analysis 79, 84
Our Common Future (Brundtland Report) 2, 23, 39
outdoor tourism activities 178
over-commodification 100–01
overtourism 103–12, 123

P&O Cruises 214–16
packaging 146, 176, 177, 182, 211–12, 218–19, 221, 250
Pact for Next Tourism Generation Skills 40, 45–46
partnerships 36, 37, 122
passion (CPD) 75–76
people (personnel) 36, 248–49
    see also education; employee attrition; employee engagement; employee satisfaction
personal development plans 76–78
personal training needs analysis 79, 84
personalization (CPD) 72–73, 75, 76
physical skills 68
pillars of sustainability 28–34, 39, 61
Plan Vivo 281
Plan Zheroes 221
planet SDGs 36
'Planet 21' (Accor) 140
Planetary Health Diet 245
planning 150
    see also management plans; personal development plans
plant-based menus 154, 236, 240, 241, 245
Planterra 279
plastics 182, 185, 190
Plastics Pact 213
plate waste 220, 221
policy development 279–80
polycrisis 312
polyester 186
population growth 232
portion control 220, 223
Portugal 175–81
pōwhiri 276
practical skills 68

pre-prepared food 220
Principles of Responsible Investing 141
prioritization (CPD) 75
PROA 302–03
problem-solving 48, 64
procedural knowledge 67
processed ingredients 236
product design 176
product lifecycle extension 186–88
production management 145–46
    see also SDG 12
professional development 55–85, 279, 280, 307, 312
professional development plans 76–78
professional learning community 76–78
promotions 78
Prop-Up Project 309
'Proposal for a Directive on Green Claims' (EU) 264–65, 270, 271, 282
purchasing (procurement) 154, 185–86, 218–19, 246, 249, 288
purpose (CPD) 72, 73, 74
purpose (organizational) 113–14
    see also strategy

quick service restaurants (QSRs) 217, 218
quick wins ('low-hanging fruit') 151, 152–53

Radiocentre 273
ReciclAção project 302–03
recycling 176, 178–79, 185, 186, 187, 190, 234, 244, 268, 278
    glass 211–12, 219, 243
recycling facilities 278
redesigning 176
reducing 177, 243
refillable dispensers 178, 182
refresher training 78
refrigeration 145, 250
refusing 176–77, 243
regional skills partnerships 45
regulation 183, 269–71, 289–90
    see also Consumer Protection from Unfair Trading Regulations (2008); Corporate Sustainability Reporting Directive
Regulation (EC) No 66/2010 265
Regulation (EU) 2017/1369 265
Reimagine Kaikōura plan 33
renewable energy 153, 155, 180, 190, 247, 249, 308
Renewable Energy Guarantees of Origin 145
repair services 178
replacing 179
reporting 134, 141–42, 143
repurposing 243, 309
reputation risk 136, 139, 142, 272
resident displacement 105

resource conservation 82
resource extraction 182
restaurants 204, 217, 218, 242–44, 277
retail (wholesale) food sector 208, 209, 212
rethinking 176, 179–80
return on investment (ROI) 147
reuse 177–78, 187, 243
rice production 203
risk management 141–42, 148
Roadmap to Net Zero 151
Royal Lancaster Hotel 181
Ryanair 273–74

Santorini 106
Scope 1 GHG emissions 144–45, 151–52, 156
Scope 2 GHG emissions 144, 145, 152, 156
Scope 3 GHG emissions 144, 145–46, 148–49, 152, 154–55
Scope 4 (avoided emissions) GHG emissions 147
SDG 4 37, 41–2, 59–60, 62, 119–20, 295, 297
SDG 8 120
SDG 10 120
SDG 11 121–22
SDG 12 42-44, 60, 122, 208, 214, 262
SDG 13 60
SDG 16 122
SDG 17 36, 37, 122
seafood 236
seasonal calendars 239
seasonal produce 236, 238–39, 241, 247, 252
'seconds' food policies 223
Sedex 187
self-awareness 48, 56, 64, 76
'Serve 360' (Marriott) 139
sharing economy 173, 174
signage 278–79
Silo 242–44
single-use items 176, 179, *190*
sink disposal units 221
skills *40*, 45, 45–46, 68
skills assessments 249
small and medium-sized enterprises (SMEs) 189
smart water management 155
social competencies 67, 68
  see also collaboration; self-awareness
social norms 222–23
social sustainability 30, 32, 68, 104, 105, 138, 140, 206, 234, 239
  see also labour practices
socially responsible investing 140
socio-emotional competencies 64
  see also normative competency; self-awareness
solar panels 153, 176, 177, *190*
sourcing 177, 187, 234–40
South Lake Tahoe 108–11
Spain 94
  Barcelona 104, 105

stakeholder engagement 137, 139, 142, 146–47, 248, 252
stakeholder interdependence 114–15, 116
stakeholder management 15, 25–26, 72, 105
stakeholder theory 113–17, 123, 125
Statistical Framework for Measuring the Sustainability of Tourism 22, *40*, 49, 134, *135*, 136, 288
stewardship principle 137, 186
Stockholm Conference (1972) 22, *39*
strategy 14–15, *48*, 56, 64, 113–14, 136
supplier management 145–46, 149, 154, 307–08
supply chain management 176, 206–08, 251–52
  see also local food
surplus food redistribution 221
sustainability culture 156
sustainability mindset 58–59
Sustainable Arctic Destination certification 115
sustainable aviation fuel 277
sustainable development, defined 2
Sustainable Development Goals (SDGs) 2, 35–44, 119–22, 288, 290, 297
  SDG 4 59–60, 62, 295
  SDG 12 60, 208, 214, 262
  SDG 13 60
Sustainable EU Tourism *40*, 46
Sustainable Travel International 275, 279, 281
systems thinking 20, *48*, 63

Takahanga Marae 32
tangible cultural heritage 91
target-measure-act approach 215
Task Force on Climate-related Financial Disclosures *135*, 141
Teaffund 221
TEAM 147
technology 106
  see also apps; automated meter readings; digital brochures; digital menus
3Rs model 175
top-down professional development 79–80
tourism 1–2, 3, 28, 90, 103, *104*
  Agenda 21 24, 39
  astrotourism 33
  audits 147–51
  carbon footprint 171, *172*
  CSR 138
  and food waste 210–11
  local food sourcing 238
  outdoor activities 178
  waste management projects 300–02
  see also cultural sustainability; EarthCheck; Green Key; Hague Declaration; Statistical Framework for Measuring the Sustainability of Tourism; Sustainable EU Tourism; Transition Pathway for Tourism; UN Tourism (UNWTO)

Index

Tourism Satellite Account 22
trading standards officers 265–66
training 78, 155, 216, 222, 249
training needs analysis 78–84
training plan design 83–84
Transition Pathway for Tourism 40, 46
transparency 137, 146–47, 148, 189, 267, 272, 280
transport sector 155, 171, 207, *213*, 235, 241
travel agencies 263–64
'Travel with Purpose' (Hilton) 139
tree planting *190*
triple bottom line (triple-P model) 14–15, 24–26, 28, *39*
    see also economic sustainability; environmental sustainability; planet SDGs
trust 136, 272, 274, 281
2030 Agenda for Sustainable Development 34–44

UK 208–12, 213–14
    see also Competition and Markets Authority; Food and Drink Pact (Courtauld Commitment); Green Claims Code
UN Tourism (UNWTO) 2, 22–23, *39*, 49, 58–59, 90, 103, 107–11, 135, 136, 171
unavoidable food waste 201
UNESCO 26–28, *39*, *40*, 44–45, 47–48, 63–66, 90, 94, 96–99, 113
    see also Convention for the Safeguarding of the Intangible Cultural Heritage; Convention on the Protection and Promotion of the Diversity of Cultural Expressions; Universal Declaration on Cultural Diversity
UNICEF 221
uniforms 186–87, 304–05
United Nations (UN) 140, 141, 245, 262
    see also Sustainable Development Goals (SDGs); UNESCO; UNICEF
United Nations Conference on Environment and Development (UNCED) 24, *39*
United Nations Environment Programme (UNEP) 22, *39*, 171–72, 177
Universal Declaration on Cultural Diversity 27–28, *39*, 90, 92, 97
universality 35

urban infrastructure 105–06, 121–22
USA 202, 261, 270, *271*

value creation 68, 114
values 68, 71, 94, 251
values-based management 113, 115–17, 124
variable air volume systems 153
vegetable production 203, *245*
vegetarian dishes 240–41, 247, *252*
vehicle fleet management 145
Venice 104–05, 106
Verified Carbon Standard 281
Virgin Atlantic 304–05

Waste Bank Project 301–02
Waste Electrical and Electronic Equipment 146
Waste (England and Wales) Regulations (2011) 183
waste hierarchy model 183–84
waste management 146, 179, 182–84, *190*, 300–03
    food waste disposal 221
    food waste reduction strategies 212–23, 242–44, 249, 277
water-efficient features 153, 155, 176, 177, *190*, 249
water usage 146, 155, 180, 186, 207
wellbeing 296–27, 312–13
Wellness of Being® framework 296, 297, 310–13
Westerveld, Jay 261
Whale Watch Kaikōura 32
whale-watching 114
whole (retail) food sector 208, 209, 212
Wivenhoe House Hotel 13–15
World Animal Protection 275
World Cetacean Alliance 114, 275
World Tourism Organization *see* UN Tourism (UNWTO)
World Travel & Tourism Council 106–11
World Wellbeing Movement Measurement 312–13
WRAP 146, 187, 208–09, 210, 211, 213–16, 217, 281

Zen Buddhism 94

From 4 December 2025 the EU Responsible Person (GPSR) is:
eucomply oÜ, Pärnu mnt. 139b – 14, 11317 Tallinn, Estonia
www.eucompliancepartner.com